Praise for *The Manager's Path*

The Manager's Path gives the big picture perspective on what a career in engineering management looks like. Camille provides very tactical advice for each career stage. And because engineering managers have a great responsibility to their reports to learn how to manage well, you should read this book and learn how it is done.

This book is a practical guide to understanding and pursuing a career in Engineering Management.

—*Liz Crawford, Entrepreneur in Residence, Genacast Ventures; former CTO, Birchbox*

As Camille says in Chapter 5, "This book is for engineering managers. It's not a generic management book." Without hesitation I recommend this book for literally everyone who works in or around software engineering, at whatever level, whether or not you believe management is for you.

In software engineering we often treat management as something between a fate to be avoided, an obstacle, and a reward for being the loudest person in the room. Is it a surprise that most of us have experienced poor management and we struggle, as an industry, to bring managers up to a level slight better than worse-than-useless? Camille's book teaches us how to clear this bar by a considerable margin. She starts from where we all start, as a human who is being managed, and works upward from that common ground. Camille is one of the great engineering leaders in our industry. Her advice is both practical and profound. While I wish I'd had this book earlier in my career, I'm grateful to have it now.

—*Kellan Elliot-McCrea, SVP Engineering, Blink Health; former CTO, Etsy*

I've learned more from Camille about engineering leadership than almost anyone. Her writing is a fantastic help to both new and experienced managers, thinking through not just how to get the job done, but how to find the best approach for both the business and the people. This will be a book I recommend to all managers for years to come.

—*Marc Hedlund, CEO, Skyliner;*
former VP Engineering at Stripe and Etsy

The Manager's Path

A Guide for Tech Leaders Navigating Growth and Change

Camille Fournier

Beijing · Boston · Farnham · Sebastopol · Tokyo

The Manager's Path

by Camille Fournier

Printed in the United States of America.

Published by O'Reilly Media, Inc., 1005 Gravenstein Highway North, Sebastopol, CA 95472.

O'Reilly books may be purchased for educational, business, or sales promotional use. Online editions are also available for most titles (*http://oreilly.com/safari*). For more information, contact our corporate/institutional sales department: 800-998-9938 or *corporate@oreilly.com*.

Editor: Laurel Ruma	**Interior Designer:** Monica Kamsvaag
Production Editor: Kristen Brown	**Cover Designer:** Edie Freedman and Michael Oréal
Copyeditor: Rachel Monaghan	
Proofreader: Rachel Head	**Illustrator:** Rebecca Demarest
Indexer: Angela Howard	

April 2017:   First Edition

Revision History for the First Edition

2017-03-08:   First Release

See *http://oreilly.com/catalog/errata.csp?isbn=9781491973899* for release details.

978-1-491-97389-9

[LSI]

To CK

Contents

Acknowledgments

Special thanks go out to my editors, Laurel Ruma and Ashley Brown, who helped this first-time author get through her book without too many tears.

Thank you to Michael Marçal, Caitie McCaffrey, James Turnbull, Cate Huston, Marc Hedlund, Pete Miron, bethanye Blount, and Lara Hogan for providing anecdotes on leadership to share with our readers.

Thanks to everyone who gave me valuable feedback during the writing process, including Timothy Danford, Rod Begbie, Liz Crawford, Cate Huston, James Turnbull, Julie Steele, Marilyn Cole, Katherine Styer, and Adrian Howard.

Special thanks to my collaborator, Kellan Elliott-McCrea, for his numerous bits of management wisdom, and to all of my CTO Dinner friends for your advice over the years, much of which made it into this manuscript.

To my long-time coach, Dani Rukin, thank you for helping me get out of my head, and for encouraging me to always stay curious.

Last but not least, thanks to my husband, Chris, for the many dinner-table debates that shaped some of the trickiest writing. His insights and edits have helped me become the writer I am today.

Introduction

In 2011, I joined a small startup called Rent the Runway. It was a radical departure for me to go from working on large distributed systems at a big company to working with a tiny engineering team with a focus on delivering a great customer experience. I did it because I thought the business was brilliant, and I wanted a chance to lead. I believed that with a little luck and some hard work, I could get that leadership experience that I was so eager to have.

I had no idea what I was getting myself into. I joined Rent the Runway as a manager without a team, a director of engineering in name and something closer to a tech lead in practice. As is often the case with startup life, I was hired to make big things happen, and had to figure out myself what that might look like.

Over the next four years, my role grew from managing a small team to running all of engineering as CTO. As the organization scaled, so did I. I had mentors, coaches, and friends who provided valuable advice, but no one was there to tell me specifically what to do. There was no safety net, and the learning curve was brutal.

When I left the company, I found myself bursting with advice. I also wanted a creative outlet, so I decided to participate in "National Novel Writing Month," which is a challenge to write 50,000 words in 30 days. I attempted to write down everything I had learned over the past four years, everything I had personally experienced and several observations I'd made watching others succeed and struggle. That project turned into the book you are reading now.

This book is structured to follow the stages of a typical career path for an engineer who ends up becoming a manager. From the first steps as a mentor to the challenges of senior leadership, I have tried to highlight the main themes and lessons that you typically learn at each step along the way. No book can cover every detail, but my goal is to help you focus on each level individually, instead of overwhelming you with details about challenges that are irrelevant to your current situation.

In my experience, most of the challenge of engineering management is in the intersection of "engineering" and "management." The people side is hard, and I don't want to sell the challenges of those interpersonal relationships short. But those people-specific management skills also translate across industries and jobs. If you are interested in improving on purely the people management side of leadership, books like *First, Break All the Rules*[1] are excellent references.

What engineering managers do, though, is not pure people management. We are managing groups of technical people, and most of us come into the role from a position of hands-on expertise. I wouldn't recommend trying to do it any other way! Hands-on expertise is what gives you credibility and what helps you make decisions and lead your team effectively. There are many parts of this book dedicated to the particular challenges of management as a technical discipline.

Engineering management is hard, but there are strategies for approaching it that can help make it easier. I hope that in reading this you get some new ideas for how you might approach the role of engineering management, whether you're just starting out or have been doing it for years.

How to Read This Book

This book is separated into chapters that cover increasing levels of management complexity. The first chapter describes the basics of how to be managed, and what to expect from a manager. The next two chapters cover mentoring and being a tech lead, which are both critical steps on the management path. For the experienced manager, these chapters have some notes on how you might approach managing people in these roles. The following four chapters talk about people management, team management, management of multiple teams, and managing managers. The last chapter on the management path, Chapter 8, is all about senior leadership.

For the beginning manager, it may be enough to read the first three or four chapters for now and skim the rest, returning when you start to face those challenges. For the experienced manager, you may prefer to focus on the chapters around the level that you're currently struggling with.

---

[1] Marcus Buckingham and Curt Coffman, *First, Break All the Rules: What the World's Greatest Managers Do Differently* (New York: Simon & Schuster, 1999).

Interspersed throughout are sections with three recurring themes:

Ask the CTO
> These are brief interludes to discuss a specific issue that tends to come up at each of the various levels.

Good Manager, Bad Manager
> These sections cover common dysfunctions of engineering managers, and provide some strategies for identifying these bad habits and overcoming them. Each section is placed in the chapter/level that is most likely to correspond to the dysfunction, but these dysfunctions are often seen at every level of experience.

Challenging Situations
> Starting in Chapter 4, I take some time to discuss challenging situations that might come up. Again, while these are roughly placed with the level that is most appropriate, you may find useful information in them regardless of your current level.

Chapter 9 is a bit of a wildcard, aimed at those trying to set up, change, or improve the culture of their team. While it was written from a perspective of a startup leader, I think that much of it will apply to those coming into new companies or running teams that need an uplift in their culture and processes.

More than an inspirational leadership book for a general-purpose audience, I wanted to write something worthy of the O'Reilly imprint, something you can refer back to over time in the same way you might refer to *Programming Perl*. Therefore, think of this book as a reference manual for engineering managers, a book focused on practical tips that I hope will be useful to you throughout your management career.

O'Reilly Safari

 *Safari* (formerly Safari Books Online) is a membership-based training and reference platform for enterprise, government, educators, and individuals.

Members have access to thousands of books, training videos, Learning Paths, interactive tutorials, and curated playlists from over 250 publishers, including O'Reilly Media, Harvard Business Review, Prentice Hall Professional, Addison-Wesley Professional, Microsoft Press, Sams, Que, Peachpit Press, Adobe, Focal Press, Cisco Press, John Wiley & Sons, Syngress, Morgan Kaufmann, IBM

Redbooks, Packt, Adobe Press, FT Press, Apress, Manning, New Riders, McGraw-Hill, Jones & Bartlett, and Course Technology, among others.

For more information, please visit *http://oreilly.com/safari*.

How to Contact Us

Please address comments and questions concerning this book to the publisher:

> O'Reilly Media, Inc.
> 1005 Gravenstein Highway North
> Sebastopol, CA 95472
> 800-998-9938 (in the United States or Canada)
> 707-829-0515 (international or local)
> 707-829-0104 (fax)

We have a web page for this book, where we list errata, examples, and any additional information. You can access this page at *http://bit.ly/the-managers-path*.

To comment or ask technical questions about this book, send email to *book-questions@oreilly.com*.

For more information about our books, courses, conferences, and news, see our website at *http://www.oreilly.com*.

Find us on Facebook: *http://facebook.com/oreilly*

Follow us on Twitter: *http://twitter.com/oreillymedia*

Watch us on YouTube: *http://www.youtube.com/oreillymedia*

Management 101

The secret of managing is keeping the people who hate you away from the ones who haven't made up their minds.

—CASEY STENGEL

You're reading this book because you want to be a good manager, but do you even know what one looks like? Have you ever had a good manager? If someone were to sit you down and ask you what you should expect from a good manager, could you answer that question?

What to Expect from a Manager

Everyone's very first experience of management is on the other side of the table, and the experience of being managed is the foundation on which you build your own management philosophy. Unfortunately, I've come to see that there are people who have never in their careers had a good manager. Friends of mine talk about their best managers as managing them with "benign neglect." The engineer just kind of knows what to work on, and the manager just leaves them alone entirely. In the most extreme case, one person reported meeting only twice with his manager in the span of six months, one of those times to receive a promotion.

Benign neglect isn't so bad when you consider some of the alternatives. There are the neglectful managers who ignore you when you need help and brush your concerns aside, who avoid meeting with you and who never give you feedback, only to tell you suddenly that you are not meeting expectations or not qualified to be promoted. And of course there are micromanagers who question every detail of everything you do and refuse to let you make any decisions on your own. Still worse are actively abusive managers who neglect you until they want to yell at you for something. Sadly, all of these characters are walking around our companies, wreaking havoc on the mental health of their teams. When you believe that these are the only alternatives, a manager who leaves you alone most of the time unless you specifically ask for help doesn't seem so bad at all.

There are, however, other options. Managers who care about you as a person, and who actively work to help you grow in your career. Managers who teach you important skills and give you valuable feedback. Managers who help you navigate difficult situations, who help you figure out what you need to learn. Managers who want you to take their job someday. And most importantly, managers who help you understand what is important to focus on, and enable you to have that focus.

At a minimum, there are a few tasks that you should expect your manager to perform as needed, in order to keep you and your team on track. As you learn what to expect from your manager, you can start to ask for what you need.

ONE-ON-ONE MEETINGS

One-on-one meetings (1-1s) with your direct manager are an essential feature of a good working relationship. However, many managers neglect these meetings, or make them feel like a waste of your time. What does it feel like to be on the receiving end of a good 1-1?

1-1s serve two purposes. First, they create human connection between you and your manager. That doesn't mean you spend the whole time talking about your hobbies or families or making small talk about the weekend. But letting your manager into your life a little bit is important, because when there are stressful things happening (a death in the family, a new child, a breakup, housing woes), it will be much easier to ask your manager for time off or tell him what you need if he has context on you as a person. Great managers notice when your normal energy level changes, and will hopefully care enough to ask you about it.

I am not a buddy-buddy person at work. I feel the need to say this because I think that sometimes we give ourselves a pass at caring about our colleagues because we're introverts, or we don't want to make friends at work. You might think that I am the sort who loves to make lots of work friends, and therefore I don't understand how this feels to you, but I assure you: I understand that you don't feel like that human side is all that interesting in the workplace. Being an introvert is not an excuse for making no effort to treat people like real human beings, however. The bedrock of strong teams is human connection, which leads to trust. And trust, real trust, requires the ability and willingness to be vulnerable in front of each other. So, your manager will hopefully treat you like a human who has a life outside of work, and spend a few minutes talking about that life when you meet.

The second purpose of a 1-1 is a regular opportunity for you to speak privately with your manager about whatever needs discussing. You should expect your 1-1s to be scheduled with some predictability so that you can plan for them, because it is not your manager's job to completely control the 1-1 agenda. Sometimes he will, but it is good for you to put a little thought into what you might actually want to discuss before your 1-1 meetings. It is hard to do this if your manager does not regularly meet with you, or constantly cancels or changes your 1-1s. You may not want to do 1-1s regularly, or you may only need them every few weeks. That's OK, so long as you don't eliminate them completely. Use them as you need them, and if you find that you want to meet more frequently, ask your manager for that.

For most people, good 1-1s are not status meetings. If you are a manager reporting up to senior management, you may use your 1-1 to discuss the status of critical projects, or projects that are still in the nascent stage where there's not necessarily a lot written down yet. If you're an individual contributor, though, a 1-1 as a status meeting is repetitive and probably boring. If your 1-1 is a dreadful obligation for delivering a boring status report, try using email or chat for that purpose instead to free up the time, and bringing some topics of your own to the 1-1.

I encourage you to share the responsibility of having good 1-1s with your manager. Come with an agenda of things you would like to discuss. Prepare for the time yourself. If he cancels or reschedules on you regularly, push him to find a time that is more stable, and if this isn't possible, verify the day before (or that morning, for an afternoon meeting) that you will be meeting and share with him anything you are interested in discussing so he knows you want to meet.

FEEDBACK AND WORKPLACE GUIDANCE

The second thing to expect from your manager is feedback. I'm not just talking about performance reviews, although that is part of it. Inevitably, you will screw up in some fashion, and if your manager is any good she will let you know quickly that you did. This is going to be uncomfortable! In particular, for those new to the workforce who are not used to getting behavioral feedback from anyone but their parents, this can be a pretty disorienting thing to have happen.

You do want to get this feedback, though, because the only thing worse than getting behavioral feedback is not getting it at all, or getting it only during your performance review. The sooner you know about your bad habits, the easier they are to correct. This also goes for getting praise. A great manager will notice some of the little things you're doing well in your day-to-day, and recognize you for them. Keep track of this feedback, good and bad, and use it when you write your self-review for the year.

Ideally, the feedback you get from your manager will be somewhat public if it's praise, and private if it's criticism. If your manager grabs you immediately after a meeting to provide critical feedback, that is not necessarily a sign that your behavior was terrible. Good managers know that delivering feedback quickly is more valuable than waiting for a convenient time to say something. Praising in public is considered to be a best practice because it helps the manager let everyone know that someone has done something laudable, and reinforces what positive behavior looks like. If you don't like public praise, tell your manager! It would be great if she asked, but if she doesn't, you shouldn't suffer in silence.

There are other types of feedback that you may want to ask for from your manager. If you are giving a presentation, you can ask her to review the content and suggest changes. If you've written a design doc, she should be able to provide ideas of areas for improvement. As engineers, we get code feedback mostly from our peers, but you will do things other than code, and your manager should act as a resource to help you improve those things. Asking your manager for advice is also a good way to show that you respect her. People like to feel helpful, and managers are not immune to this sort of flattery.

When it comes to your role at the company, your manager needs to be your number one ally. If you're at a company with a career ladder, sitting down with your manager and asking her what areas you need to focus on to get promoted is usually a good idea if you are actively seeking a promotion. If you're struggling with a teammate or a person on another team, your manager should be there to help you navigate that situation, and she can work with the other person or team as necessary to help you get to resolution. This usually requires you to say something, though. If you don't ask your manager about a promotion, do not expect her to just give you one magically. If you're unhappy with a teammate, your manager may not do anything unless you bring the issue to her attention.

It's great when managers can identify and assign stretch projects that will help us grow and learn new things. Beyond assigning stretch projects, though, good managers will also help you understand the value of the work you're doing even when it is not fun or glamorous. Your manager should be the person who shows you the larger picture of how your work fits into the team's goals, and helps you feel a sense of purpose in the day-to-day work. The most mundane work can turn into a source of pride when you understand how it contributes to the overall success of the company.

As you become more senior, the amount of personal feedback you get, both good and bad, is likely to decrease. You are operating at a higher level, and your manager is operating at a very high level. Expect the type of feedback to change somewhat from personal feedback to team- or strategy-related input. It's even more important as you become more senior that you feel comfortable driving your 1-1s and bringing topics for discussion or feedback to your manager, because she is otherwise unlikely to spend a lot of time on this outside of performance reviews.

TRAINING AND CAREER GROWTH

As the main liaison between you and the bureaucracy of the company, your manager holds some responsibility for helping you find training and other resources for career growth. This may be helping you find a conference to attend or a class to take, helping you get a book you need, or pointing you to an expert somewhere else in the company who can help you learn something.

The role of manager as the person who provides mentoring and training is not a universal expectation. In some companies, these areas are entirely managed by a training arm that you can tap into directly. Some companies are too small to have the money to provide much training, or don't think it's a necessary perk to offer employees.

Whatever kind of company you work for, expect that you are responsible, for the most part, for figuring out what types of training you want. This is especially true for individual contributors looking for training in technical areas. Your manager is unlikely to just have a list of interesting conferences or training opportunities at his fingertips.

The other way your manager will contribute very directly to your career growth is via promotion and, probably, compensation. If your company has a promotion process, your manager will be involved in it in some fashion. For companies that do promotion via committee, your manager will guide you through the process of preparing your promotion packet—the set of materials that the committee will review. If your manager or the management hierarchy determines promotions directly, your direct manager will be essential in advocating for your promotion and getting it approved.

In whatever way promotions happen, your manager should have an idea of whether you are qualified to be promoted. When you are interested in being promoted, it's very important to ask your manager for specific areas to focus on in order to get that promotion. Managers usually cannot guarantee promotions, but good managers know what the system is looking for and can help you build those achievements and skills. Again, this only goes so far. At more senior levels of work, opportunities for promotion are much more rare, and your manager may need you to find and propose the achievements that qualify you for the next level.

Ask the CTO: Big Ambitions

I've just started working, but I already know that my career goal is to someday become a CTO myself. What should I be doing now to make that possible?

The very first thing you need to learn is how to work. Maybe you know this already, but when I was first out of college, I didn't really know how to do the job at all. Because the day-to-day work of tech is very different from school, there are probably a bunch of things you will learn about the process of being a working engineer. My specific advice would be to seek out a workplace where you can get mentorship and training in the aspects of doing the job (such as testing, project and product management, and collaboration) as well as learn new technical skills. You want to build a strong foundation of skills because you will need them to succeed.

I also advise you to find the best managers and mentors you can, and watch them work. Try to find people to work for who push you to succeed but also reward you for success, who inspire you to stretch yourself. Realize that stretching yourself is about more than just learning new technologies: great CTOs have strong communication skills, project management skills, and product sense, in addition to good technical sense. However, you also want to spend a lot of time writing code, and getting really good at understanding how high-quality code is written. This will probably take a few years of focus, and you can't rush it.

In addition, I encourage you to create and build a strong network of peers. One thing that early career engineers often don't appreciate is how their current peers will turn into their future jobs. This peer group includes everyone from your schoolmates to your teammates to the people you meet at conferences and meetups. It's OK to be a little bit shy, but most CTOs have to learn how to socialize with all sorts of people and create strong networks across companies.

One final thing to realize is that most CTOs are the CTOs of small companies. They are often the technical cofounders of startups. If you want to go down that path, the best thing to do is to work for a company that has a track record of people who have left it to found new companies. This is where you will meet your future cofounders and discover opportunities to get into new companies early.

How to Be Managed

Part of being a good manager is figuring out how to be managed. This is not exactly the same as managing up, although it is related. Developing a sense of ownership and authority for your own experiences at work, and not relying on your manager to set the entire tone for your relationship, is an important step in owning your career and workplace happiness.

SPEND TIME THINKING ABOUT WHAT YOU WANT

Your manager can point out opportunities for growth. She can show you projects. She can provide feedback on your areas of learning and development. But she cannot read your mind, and she cannot tell you what will make you happy. Whether you are brand new to the workplace or 20 years into your career, the onus of figuring out what you want to do, what you want to learn, and what will make you happy rests on *your* shoulders.

You'll probably go through periods of career uncertainty in your life. Many people feel very uncertain in their first two to five years out of school, as they settle into independent adulthood. I felt so unsettled that I went to grad school for a couple of years, in what turned out to be a quest to find security in the familiar academic setting and an escape from a job I didn't know how to navigate well. I hit uncertainty again after climbing the technical ladder only to feel somewhat powerless at a big company. And then I hit it again after climbing the management ladder only to encounter the challenges of executive leadership. I expect I'll experience it every 5 to 10 years until I retire, given my track record.

As you go through various stages of your career, you'll start to realize how much uncertainty there is in the world. It's a pretty universal truth that once you get the job you thought you wanted, the enjoyment eventually fades and you find yourself looking for something else. You think you want to work for that cool startup, and you get there only to find it's a mess. You think you want to be a manager, only to discover that the job is hard and not rewarding in the ways you expected.

In all of this uncertainty, the only person you can rely on to pull through it is yourself. Your manager cannot do that for you. Use your manager to discover what's possible where you are, but look to understand yourself in order to figure out where you want to go next.

YOU ARE RESPONSIBLE FOR YOURSELF

Knowing yourself is step one. Step two is going after what you want.

Bring agendas to your 1-1s when you have things you need to talk about. When you want to work on projects, ask. Advocate for yourself. When your manager isn't helpful, look for other places to get help. Seek out feedback, including constructive feedback on areas to improve. When that feedback comes to you, take it graciously, even when you don't agree with it.

When you are persistently unhappy, say something. When you are stuck, ask for help. When you want a raise, ask for it. When you want a promotion, find out what you need to do to get it.

Your manager cannot force work–life balance on you. If you want to go home, figure out how to get your work done and go home. Sometimes you will have to go against the cultural grain to set your own boundaries, and that will feel uncomfortable. On the flip side, sometimes if you want a bigger job, you will have to work more hours to get it.

You will not get everything you ask for, and asking is not usually a fun or comfortable experience. However, it's the fastest way forward. If your manager is conscientious, he'll appreciate your candor. He may not be conscientious, or he may like you less for asking, and then you'll know that about your current situation. I can't guarantee you that it'll go well, but if you've set a goal for yourself, you owe it to yourself to do what you can to make it happen.

GIVE YOUR MANAGER A BREAK

This is a job. Your manager will be stressed out sometimes. She'll be imperfect. She will say dumb things, or do things that feel unfair or harmful to you. She'll give you work that you don't want to do, and get annoyed when you complain about doing it. Her job is to do the best thing for the company and the team. It is not to do whatever it takes to make you happy all the time.

Your relationship with your manager is like any other close interpersonal relationship. The only person you can change is yourself. You should absolutely provide feedback to your manager, but understand that she may not listen or change no matter how much you think she should. If you find yourself starting to actively resent your manager for whatever reason, you probably need to move to a different team or look for a new job. If you find yourself resenting *every* manager you work for, you may need to think about whether the cause is them or you. Perhaps you'd be happier in a job where you don't have a manager.

Especially as you become more senior, remember that your manager expects you to bring solutions, not problems. Try not to make every 1-1 about how you need something, how something is wrong, or how you want something more. When you have a problem, instead of demanding that your manager solve it for you, try asking her for advice on how she might approach the problem. Asking for advice is always a good way to show respect and trust.

CHOOSE YOUR MANAGERS WISELY

Your manager can make a huge difference in your career. So, as much as you can, consider not only the job, the company, and the pay, but also the manager when you are evaluating job opportunities.

Strong managers know how to play the game at their company. They can get you promoted; they can get you attention and feedback from important people. Strong managers have strong networks, and they can get you jobs even after you stop working for them.

There's a difference between a strong manager and a manager that you like as a friend, or even one you respect as an engineer. Plenty of great engineers make ineffective managers because they don't know or want to deal with the politics of leadership in their companies. A strong engineer may make a great mentor-manager to someone early in his career, but a terrible advocate-manager for someone who is more senior.

Assessing Your Own Experience

Here are some questions to consider as you develop this part of your career:

- Have you had a manager you considered good? What did this manager do that you found valuable?

- How often do you meet 1-1 with your manager? Do you come to 1-1s with your manager bringing topics to discuss? If your 1-1 is a status meeting, can you use some other means to convey that status?

- Do you feel that you can tell your manager when you have a major life event? Do you feel that your manager knows something about you personally?

- Has your manager delivered good feedback to you? Bad feedback? Any feedback at all?

- Has your manager helped you set any work-related goals for this year?

2

Mentoring

The first act of people management for many engineers is often unofficial. They find themselves, through the luck of the draw, mentoring someone.

The Importance of Mentoring to Junior Team Members

Mentors are commonly assigned to junior members of a team, such as new hires straight out of school or student interns. Many organizations use mentors as part of their onboarding process for all new hires. Sometimes the mentor is another junior person on the team, perhaps herself only a year or two into the organization; someone who can still clearly remember the onboarding or internship process herself, and can closely relate to the new person. Other times the mentor is a senior engineer who can act as a technical mentor in addition to helping the new hire get up to speed on the process. In a healthy organization, this onboarding mentorship role is used as an opportunity for both parties. The mentor gets the chance to see what it is like to have responsibility for another person, and the mentee gets an overseer who is focused on him alone, without other reports clamoring for his mentor's attention.

I remember my first mentor, who guided me through my first serious taste of working as a software engineer. I was an intern at Sun Microsystems, working on a team that wrote JVM tools. This was the first job where I had a real software project to build, and I was lucky enough to have a great mentor, a senior engineer named Kevin. Kevin was a memorable mentor because, despite being a senior technical leader in the area we were working in, he made time for me. Instead of showing me a desk and leaving me to figure out what exactly I needed to do, Kevin took the time to discuss projects with me, to sit with me at the whiteboard, to go through the code together. I knew what I was expected to get done, and when I got stuck, I could ask him for help. That summer was critical for my development as a software engineer, because under his guidance I began to see that I could actually do real-world work and that I was capable of being a productive employee. Working with Kevin was one of my first major career milestones. This experience taught me the value of mentorship.

Being a Mentor

If you find yourself in the mentor's seat, congratulations! This is an experience that not everyone will get: an opportunity to learn in a fairly safe way about the job of management, and the feeling of being responsible for another person. It's unlikely that you'll get fired for being a bad mentor (unless, of course, you behave in an inappropriate manner—please don't hit on your mentee!). For many mentors, the worst that can happen is that a) the mentee is a drain on their time and they get less coding work done, or b) they do such a poor job that someone whom the organization might otherwise want to hire/keep around has a bad experience and doesn't join the organization, or opts to leave the organization sooner than she otherwise might. Sadly, the second outcome is far more likely than the first. Great talent is sometimes squandered by weak mentors who do little but ignore their charges, waste their time with trivial projects, or, worst of all, intimidate and belittle them out of ever wanting to join the organization. But you, dear reader, don't want to do this. You want to be a great mentor! Or perhaps you are already a manager, looking to make your team more effective at the mentoring relationships you need them to take on. How do you create good, effective mentoring relationships without slowing development down too much?

MENTORING AN INTERN

The first type of mentorship relationship we'll cover is the temporary employee. For most tech companies, this is a summer intern, some bright student still in the midst of earning a degree and looking to get some valuable experience by working for your company. The screening process for such students varies; many companies view these opportunities as a pipeline to hiring great talent straight out of college, but if you're taking on someone who is more than a year from graduation, it's more realistic to expect that the candidate will a) know very little, and b) probably go elsewhere next year for his internship unless he has an amazing experience. *No pressure.*

So you find yourself mentoring a college student with little real experience. How can you make sure that his summer is awesome? Even if your company doesn't love him, you want him to love you, because he'll go back and tell all his friends about the summer he had working for your company. That can have a big impact on your ability to hire full-time from the graduating class, and the fact that you pulled interns from that school probably indicates a serious interest in hiring new graduates full-time as well. But don't worry! Making interns happy isn't rocket science.

The first thing you need is some sort of project for this intern to work on. It would be nice if you, as the mentor, weren't stuck coming up with the idea for this project, because doing so can be a daunting task. Without a project, your intern has a good chance of being completely lost and bored the entire summer. Figuring out what to do in a workplace is hard enough for experienced hires, so it's an especially tall order for an intern. You have to have a project in mind—at least something to get him started for the first couple of weeks. If you're truly drawing a blank, look at small features of your own current project that would take you a few days to complete, and start there.

The intern's first few days will be similar to those of any new hire: onboarding, getting used to the office, meeting people, learning the systems. Sit with him as much as possible these first few days. Get him started installing the IDE and checking out the code. Touch base several times a day to make sure he's not feeling lost or overwhelmed by the volume of new information. In the meantime, prepare yourself for his project.

Once you have a project, start applying your budding knowledge of project management to the task at hand. Is this project broken down into milestones? If not, spend a little time in the first few days of the intern's tenure breaking it down. Walk through the breakdown with your intern. Does it make sense to him? Listen to his questions and answer them. Remember, you're practicing skills that you will need should you decide in the future to become a manager. In this case, these skills are listening, communicating what needs to happen, and adjusting to his responses.

Listen carefully

Listening is the first and most basic skill of managing people. Listening is a precursor to empathy, which is one of the core skills of a quality manager. You need this skill wherever your career takes you; even principal engineers with no reports need to be able to hear what others are really saying. So, when your mentee is speaking to you, pay attention to your own behavior. Are you spending all your time thinking about what you want to say next? Are you thinking about your own work? Are you doing anything other than listening to the words coming out of his mouth? If so, you're not listening well.

One of the early lessons in leadership, whether it is via direct management or indirect influence, is that people are not good at saying precisely what they mean in a way that others can exactly understand. We have yet to achieve Borg hive mind or Vulcan mind meld, so we're constantly pushing complex ideas through the eye of the needle of language. And language is not something that most engineers have mastered in nuance and interpretation. So listening goes beyond hearing the words your mentee is saying. When you're face to face with another person, you also have to interpret his body language and the way he's saying those words. Is he looking you in the eye? Is he smiling? Frowning? Sighing? These small signals give you a clue as to whether he feels understood or not.

Be prepared to say anything complex a few times, in different ways. If you feel that you don't understand something your mentee has asked you, repeat the question in a different way. Let him correct you. Use those whiteboards scattered around your office, if necessary, to draw diagrams. Spend the time that you need to spend to feel understood, and like you understand the mentee. And remember that you're in a position of huge power in your mentee's eyes. He's probably nervous about screwing this opportunity up, trying his best to please you, and trying hard not to look stupid. He may not ask questions even when he doesn't understand things. Make your life easier and get those questions out of him. The odds of you spending all of your time answering questions are slim compared to the odds that your intern will go off in the absolute wrong direction because he didn't ask enough questions.

Clearly communicate

With that said, what if the intern does spend too much time asking you for help, without ever looking for help himself? Well, that gives you an opportunity to work on another management skill: communicating what needs to happen. If you expect him to do research on his own before asking you a question, tell him so! Ask him to explain a piece of code to you, or some product or process, and point him to the documents that you believe explain it. If he can't do it even with pointers, well, you're starting to learn something about the potential of this intern. If all else fails, give him the first milestone of the project and tell him to work on it alone for a day or two. Therein lies the value of breaking the project down before the intern starts working on it: you've taken on some of the harder thinking up front. He may surprise you by finishing everything much faster than you anticipated, but what a happy surprise to have! Generally, you'll need to provide some nudging and clarity along the way to keep the intern going in the right direction.

Calibrate your response

This brings us to the final management skill for you to practice: adjusting to the intern's responses. So many things can happen in the course of this mentoring relationship. He can far outstrip your expectations. He can struggle with simple tasks. He can produce work very quickly, but it's of poor quality, or he can work very slowly to produce something that's overly perfect. In the first few weeks of the internship, you're learning the frequency that you need to check in with him to provide the right adjustments. It may be once a week. It may be once a day. It may be less frequently than once a week, but I would recommend trying to check in once a week regardless, and spending any extra time as an additional interview/sell cycle for the company.

Hopefully, the summer ends on good terms. He completed a project that has some value. You got practice listening, communicating, and adjusting. He leaves thinking happy thoughts about your company, and you have gained some insight as to whether you want to do this management thing now, soon, or ever. Congratulations!

Ask the CTO: Mentoring a Summer Intern

I'm supposed to mentor a summer intern, but I have no idea where to begin. What should the intern do? How do I prepare to help this person have a great summer?

Preparing for a summer intern shouldn't take much time, but it's crucial to your success in mentoring him. Here are the basic things you need to do:

1. **Prepare for his arrival.** Do you know what day he arrives? If not, find out. Then make sure that the physical and digital environment will be set up for him when he gets there. Will he have a desk near you? A computer? Access to the systems and software? Even at big companies, a lot of these steps are sometimes overlooked for interns, and there's nothing worse than showing up for your big job with nowhere to sit and no access to the systems.

2. **Have a project for him to work on.** The best internships are those with a clear project. The trick to deciding on an intern project is that you want something specific but not urgent, something relevant to the team but also something that could be completed by an entry-level engineer in the span of, say, half the time that your intern will be there. So if he'll be there for 10 weeks, give him a project you think would probably take a new hire about 5 weeks. This achieves two goals. It gives him plenty of time, so if he has a lot of other activities to do, such as going to training or social events set up by your internship program, he will still have time to complete it. If he completes it before the end of the program, great—he should hopefully know enough about some part of your code base to do other work for the rest of the time. Remember, this person is an intern. He's still in school and still learning, so expect him to move slowly, and be pleasantly surprised if he overshoots.

3. **Plan to have him present the work he did at the end of the program.** This helps him get exposure beyond you and the other mentors, and gives him the clear expectation that you want him to finish a project. It's likely that you'll be a big part of the decision about whether or not your company makes a full-time offer to this intern, or brings him back for another summer if he isn't graduating yet. You'll probably need to spend some time coaching him on how to make the presentation. If your team does regular demo days or team meetings, the presentation should probably follow that format. There's no need for the presentation to be long or detailed, but showing off his work to the team is a great way to help your intern feel like his work mattered. I promise you that interns who feel like the company appreciated their work are the ones most likely to come back after they graduate.

MENTORING A NEW HIRE

My first job out of college was at a very large tech company. We'll call it Big-TechCo. I was put on a team that was in the midst of releasing a project several years in the making. My manager showed me to my office and then left me alone to figure out for myself what needed to be done. I didn't know how to ask for help, and I was afraid that I would be seen as a fool if I did. Unsurprisingly, I got discouraged, and in my discouragement I decided that the best thing to do was to go to graduate school. So I did.

My first job out of graduate school could not have been more different. Instead of being shown a desk and left alone, I was set up with a mentor. He encouraged me to ask questions. We did some pair programming so that I could learn the code base, and the way that testing worked for this project (my first taste of unit testing!). I was productive within days, and learned more in the first few months of that job than I had learned in the entire time I worked at Big-TechCo. I credit this almost entirely to the mentoring I got when I started.

Mentoring new hires is critical. Your job as a new hire mentor consists of onboarding, helping this person adjust to life in the company effectively, and building your and her network of contacts in the company. It can be an easier job than mentoring an intern, but the relationship and mentoring will usually go on for a lot longer.

This is an opportunity for you to see the world of your company through fresh eyes. It can be hard to remember what it was like to experience your world for the first time. How does work get done? What are the rules, spoken and unspoken? For example, you may have a standard vacation policy in the HR handbook; this is a spoken rule. The unspoken rule is that you don't take vacation the week after Thanksgiving because you're in ecommerce and that's an important week for the business. A more subtle unspoken rule dictates approximately how long you are expected to struggle with something by yourself before asking someone else to help you. There are many bits of process, culture, and jargon that are so second-nature that you might not realize they're completely foreign to a newcomer. Noticing these things gives you the opportunity to clarify them. Unspoken rules don't just make it harder for new people to join, they can also make it harder for you to do your job well. So take full advantage of this gift of fresh perspective.

Effective teams have good onboarding documents they provide to new hires. Things like step-by-step guides to setting up their development environments, learning how tracking systems work, and familiarizing themselves with the tools they will need for the job are crucial for new hires. These documents should constantly evolve to meet the changes of the workplace itself. Mentoring a new hire by helping her work through the documents, and having her modify those documents with any surprises she encounters during onboarding, provides a powerful message of commitment to her. It shows her that she has the power and obligation to learn, and to share what she's learned for the benefit of your whole team.

Part of the mentoring opportunity here is the chance to introduce the new person around. Companies are full of human networks that exist to transmit knowledge and information quickly. Bringing this person into some of your networks will help her get up to speed faster, and it will give you a new entree into whatever networks she ends up forming and joining in her time with the company. People planning on staying with the same company for a long time, particularly in large companies, often find opportunities via informal networks. Your mentee may someday be on a team that you are interested in joining, or you may someday want to bring her into a team you are running in another area.

Even if you have absolutely no interest in management, it's very difficult to build a career at any company with multiple teams without building a strong network of trusted people to share information and ideas with. The workplace is built around humans and their interactions, and these networks form the basis of any career, whether it's focused around management or individual technical contributions. You may be an introvert, or someone who does not find socializing easy, but conscious effort and practice in getting to know new people and helping them succeed will pay off. Your attitude about this will determine success or failure. Adopt the mindset that network building is a worthwhile investment of your time and energy.

TECHNICAL OR CAREER MENTORING

I'm only going to say a few words on this topic, because this type of mentoring is usually not directly related to the path of management. With that said, at some point in our careers most of us will engage in some degree of technical mentoring, career mentoring, or both. Many of us will also be given a mentor, or perhaps be encouraged to find a mentor. How can you make this type of mentoring effective?

The best mentoring relationships evolve naturally and in the context of larger work. When a senior engineer mentors a junior engineer on the team in order to help him be more productive, they can work on problems together that are relevant to both of them. The senior engineer gets value because the code written by the mentee is better, requires fewer revisions, and develops faster. The junior engineer obviously gets the value of hands-on instruction and access to someone with a deep understanding of the context he is working in. This type of mentoring is usually not a formal relationship and may be an expected part of the job for senior engineers because it delivers so much value to the team.

Many companies run formal mentoring programs where they match people up across teams, and while these programs can sometimes enhance networks, they are often an ambiguous obligation for both the mentor and the mentee. If you find yourself in one of these relationships, the best thing you can do is be specific about your expectations and goals around it.

When you are a mentor

Tell your mentee what you expect from him. If you want him to come prepared for your meetings with questions he has sent you in advance, ask for that. Be explicit about your time commitment. And then be honest with him when he asks questions. There's no point in being a mentor to a relative stranger if you can't at least use that professional distance to offer him the kind of candid advice that he may not get from his manager or coworkers.

It's also OK to say no to mentoring. Sometimes you can feel obligated to say yes to every person who asks you for help, but your time is valuable. Don't do it unless you think it will be rewarding for you and the person you're mentoring. If someone asks you to be his mentor and you can't accept, it's best just to say that you can't do it. Don't feel like you must give him a reason just because he asked. When your manager asks you to mentor someone and you don't have the time to do it, saying no is trickier. You may need to give your manager some reasons, such as your current workload, a planned vacation, or other commitments that would make mentoring impossible. Whatever you do, don't say yes and then fail to actually do the mentoring work.

When you are a mentee

Think about what you want to get out of this relationship, and come prepared to your sessions. This advice is especially relevant if you're getting mentorship from someone outside of your company, who is not being paid but is volunteering as a friendly gesture. You owe it to this person not to waste her time. If you don't have the time to prepare or don't feel that preparation is necessary, ask yourself whether the mentoring relationship is really something you need at all. Sometimes we end up with mentors because someone thinks we should have them, but there are only so many people we can meet for coffee, and only so many hours in the day. You don't have to have a mentor. Maybe instead you need a friend, or a therapist, or a coach. It can be easy to undervalue your mentor's time, because you usually aren't paying for it, so be respectful and consider finding a paid professional to help you instead.

Good Manager, Bad Manager: The Alpha Geek

In some offices, whether in a mentoring relationship or outside of one, you'll encounter an "alpha geek." The alpha geek is driven to be the best engineer on the team, to always have the right answer, and to be the person who solves all the hard problems. The alpha geek values intelligence and technical skill above all other traits, and believes these attributes should determine who gets to make decisions. The alpha geek usually can't deal with dissent, and is easily threatened by those she perceives as trying to steal her spotlight or who might upstage her. She believes herself to be the best, and responds only to messages that support that view. The alpha geek tries to create a culture of excellence, but ends up creating a culture of fear.

The alpha geek is usually an excellent, effective engineer who goes into management either because she was pushed into it or because she believes that the smartest person on the team should be the manager. She tends to undermine the people who work for her by belittling their mistakes and, at her worst, redoing the work of her teammates without warning. Sometimes the alpha geek will take credit for all of the work that a team does rather than acknowledging the strength of the team members.

At their best, alpha geeks can be inspirational to younger developers, even though they seem very intimidating. He has all the answers. She worked on the original version of that system 10 years ago and still knows the authors, and if you need to figure something out, she can do it without a problem. He knows exactly why that thing you're trying to do won't work, and when it doesn't, believe me, he'll remind you how he told you so. If only you had listened to him and done things his way! Alpha geeks have a lot to teach you, if they want to, and they can design great systems that can be fun to help build. In general, alpha geeks would not have gotten as far as they have without being very smart, so they do have a lot that they can teach their teams, and many engineers respect that intelligence enough to put up with the downsides.

At their worst, alpha geeks can't let anyone else get any glory without claiming some of it for themselves. They are the origin of any good ideas but had no part in creating the bad ideas, except that he knew they would fail. The alpha geek believes that every developer should know exactly what she knows, and if you don't know something, she will gleefully point out your ignorance. The alpha geek can be very rigid about how things should be done and closed off to new ideas that he didn't come up with. Alpha geeks get very threatened when people complain about systems they built or criticize their past technical decisions. They absolutely hate it when they have to take direction from anyone they don't respect intellectually, and can be very demeaning toward people in nontechnical roles.

The alpha geek habit often starts to show up when engineers first become mentors. If you have ever wondered why people don't seem to come to you for help despite your clearly strong technical skills, ask yourself whether you're showing some signs of being an alpha geek. Do you view yourself as an engineer who does not pull any punches and always says exactly what you think? Are you eagerly seeking out the gotcha, hunting for mistakes, reluctant to admit that someone else has had a good idea or has written good code? Do you believe that correctness is so much more important than anything else that it is always worth fighting hard for what you believe to be correct?

If you suspect that you may be an alpha geek, mentoring can be a great opportunity to break out of that habit. If you view your mentee as someone to teach and guide, where your goal is to help her in the way that best works for her, you can start to see where your aggressive style makes it harder for her to learn. Practicing the art of teaching can help us learn how to nurture and coach, how to phrase things so that others will listen, instead of just shouting them down. On the flip side, if you're unwilling to change your style to help a mentee succeed, please don't volunteer to be a mentor!

Alpha geeks make absolutely terrible managers, unless they can learn to let go of their identity as the smartest person in the room and most technical person on the team. Highly technical hands-on managers can be good for small teams of senior engineers, but alpha geeks are often better off kept out of management and given more of a focus on technical strategy and system design. You tend to see alpha geeks in the CTO role at technology-focused startups, where they are given a design and development focus across from an execution-focused Vice President of Engineering.

If you're ever in the position to promote people to management, be very, very careful in giving your alpha geeks team management positions, and keep a close eye on the impact they have in that role. The alpha geek culture can be very harmful to collaboration and can deeply undermine those who feel unable to fight back. Alpha geeks who believe that their value comes from knowing more than others can also hide information in order to maintain their edge, which makes everyone on the team less effective.

Tips for the Manager of a Mentor

What you measure, you improve. As a manager you help your team succeed by creating clear, focused, measurable goals. So often, we fail to apply this basic wisdom to the process of assigning mentors, but it applies here as much as anywhere else. When you need to assign a mentor for your new hire or intern, figure out what you're hoping to achieve by creating the relationship. Then, find the person who can help meet those goals.

First of all, figure out why you are setting up this mentoring relationship in the first place. In the two cases I discussed earlier, the mentoring relationship existed for a very specific purpose: helping a new person on the team, whether a full-time new hire or someone who will only be around for a few months, get up to speed and be productive. Of course, those aren't the only kind of mentoring programs that companies run. Sometimes people set up mentoring programs to help junior people pair with senior people outside of their team, for career or skills growth. These programs can be nice, but often the mentor and mentee are given very little guidance beyond the fact that they have been matched together. Most of the time, these programs yield very little to either party. If the mentor is not engaged or is too busy to spend any time on this project, it's a disappointment for the mentee. If the mentee doesn't know how to ask for help or what to do with the mentoring relationship, it often feels like forced socializing and a waste of time for both parties. So if your company is setting up mentoring programs outside of new hires and interns, try to make sure that there is some guidance and structure to the program before you push people into it.

Secondly, recognize that this is an additional responsibility for the mentor. If the mentor does a good job, her productivity may slow down some during the mentoring period. If you've got an engineer involved in a time-sensitive project, you may not want to push him into mentoring at the same time. Because this is an additional responsibility, treat it as you would any other important additional responsibility you might hand out. Look for someone that you believe can succeed in the role, and who wants to distinguish herself beyond her coding ability.

Whatever the source of the mentoring arrangement, common mentoring pitfalls include viewing it as a low-status "emotional labor" position, assuming that "like" must mentor "like," and failing to use the opportunity to observe potential on your team firsthand.

Emotional labor is a way to think about traditionally feminine "soft skills"—that is, skills that address the emotional needs of people and teams. Because the outcome can be hard to quantitatively measure, emotional labor is often dismissed as less important work than writing software. It's assumed to be something that should just be provided without financial recognition. I'm not suggesting that you should pay people extra money to serve as mentors, but they need to be recognized for the work they put in, and the mentor should be treated as a first-class citizen with respect to other responsibilities the person might have. As I said before, plan for it, and provide the mentor the time to do the job right. You have already invested in creating this mentoring relationship, whether it's the thousands of dollars and many hours spent on hiring, or the overhead and coordination of creating a mentoring program. It's worth continuing the investment through to fruition by recognizing that mentoring is work that takes time, but also yields valuable returns in the form of better employee networks, faster onboarding, and higher internship conversion.

When I ask you to not assume that like must mentor like, I mean that you should not expect women to only mentor women, and men to only mentor men, people of color (PoC) to only mentor other PoC, and so on. This comes up a lot in mentorship programs. These types of mentoring relationships have their place, but as a woman in tech, I personally get tired of the only mentoring being focused around lines of diversity. As you're thinking about creating mentorship relationships, unless the purpose of the mentoring program is driven from a diversity focus, give people the best mentor for their situation. Like mentoring like does make sense in one case—namely, having mentors from similar job roles. When the mentoring is expected to have a job skills training component, the best mentors are going to be people who are further along in their mastery of the job skills that the mentee is trying to develop.

Finally, use this opportunity to reward and train future leaders on your team. As you know by now, leadership requires human interaction to exist. Developing patience and empathy is an important part of the career path of anyone working in a team-based environment. Brilliant, introverted developers may not ever want to formally manage, but encouraging them to mentor 1-1 helps them develop stronger external perspectives, not to mention their own networks. Conversely, an impatient young engineer may find a degree of humility when tasked with helping an intern succeed (under your supervision).

Ask the CTO: Hiring Interns

My company has had several inquiries as to whether or not we hire interns. We have not in the past, but I'm tempted to start doing this in order to increase our hiring pool. What should I be thinking about?

Internship programs are a great way for companies to increase their hiring pipeline and find strong candidates before they graduate. However, many companies think that the goal of an internship program is to hire interns who will do a lot of work for them, and thus they miss the value of the program. I do have a couple of pieces of advice:

- **Don't hire interns who are not going to graduate in the year after their internship.** These days, college graduates from technical programs have so many options, and it's unlikely that an intern you hire who is not close to graduation is going to come back and work for you full-time. Your internship program is not a way for you to get extra work done in the summer; it's a way for you to identify and attract talent. People who are two or more years away from graduation are likely to look for new opportunities to explore before they commit to their first full-time job. When you're hiring only a handful of interns, you want all of them to have high potential for becoming full-time employees.

- **Hiring interns is relatively easy compared to hiring full-time graduates.** There is simply less demand for interns, and thus you should have a lot of options. You can choose to take this opportunity in many ways, but encourage you to push for hiring candidates from underrepresented groups. Diversity in your internship program will translate to diversity in your new grad hiring, which in turn will translate to diversity in your organization.

Key Takeaways for the Mentor

It's important to focus on three actions for yourself as a mentor.

BE CURIOUS AND OPEN-MINDED

As you grow in your career, you'll experience a lot of teachable moments, a lot of lessons in how things should or should not be. These can be "best practices," or scars caused by mistakes. This unconscious buildup can cloud our thinking and reduce our creativity. When we close our minds and stop learning, we start to lose the most valuable skill for maintaining and growing a successful technical career. Technology is always changing around us, so we must continually experience that change.

Mentoring provides a great opportunity to cultivate curiosity and see the world through fresh eyes. When faced with a mentee's questions, you can start to observe what about your organization is not so obvious to a new person. You might find areas you thought you understood but cannot explain clearly. And you'll have the opportunity to review the assumptions you've collected in your time working that may be worth questioning. While many people think creativity is about seeing new things, it's also about seeing patterns that are hidden to others. It's hard to see patterns when the only data points you have are your own experiences. Working with new people who are learning things for the first time can shed light on these hidden patterns and help you make connections you may not otherwise have made.

LISTEN AND SPEAK THEIR LANGUAGE

Mentoring, when done well, starts to shape the skills every future leader needs. Even for those who won't go on to make management their career, there are clear benefits to taking some time to mentor and learn from the experience, because mentoring forces you to hone your communication skills. It requires you to practice listening, in particular, because if you can't hear the questions you're being asked, you'll never be able to provide good answers.

Senior engineers can develop bad habits, and one of the worst is the tendency to lecture and debate with anyone who does not understand them or who disagrees with what they are saying. To work successfully with a newcomer or a more junior teammate, you must be able to listen and communicate in a way that person can understand, even if you have to try several times to get it right. Software development is a team sport in most companies, and teams have to communicate effectively to get anything done.

MAKE CONNECTIONS

Your career ultimately succeeds or fails on the strength of your network. Mentoring is a great way to build this network. You never know—the person you mentor could provide the introduction to your next job, or even come work for you in the future. On the flip side, don't abuse the mentoring relationship. Whether you're in the mentor's seat or acting as the mentee, remember that your career is long and the tech world can be very small, so treat the other person well.

Assessing Your Own Experience

Here are some questions to consider as you develop this part of your career:

- Does your company have an internship program? If so, can you volunteer to mentor an intern?

- How does your company think about onboarding? Do you assign mentors to new hires? If not, can you propose to your manager that you try doing this, and volunteer to mentor someone?

- Have you ever had a great mentor? What did that person do that made you think he or she was great? How did the mentor help you learn—what did he or she teach you?

- Have you ever had a mentoring relationship that didn't work out? Why didn't it work out? What lessons about that experience can you apply to avoid similar failures going forward?

3

Tech Lead

I became a tech lead many years ago. I had been promoted to senior engineer, and was working on a small team with several other senior engineers. It was kind of a surprise that I was promoted to tech lead, because I wasn't the most senior person on the team by either title or years of experience. In retrospect, I had a few advantages. For one, I was more than just a good engineer. I was a good communicator. I could write clear documents, I could give presentations without melting down, and I could talk to people in different teams and different roles and explain what was going on. I was also good at prioritizing. I was eager to push work forward and decide what needed to be done next. Finally, I was willing to pick up the pieces and do what needed to be done to make progress. I think, in the end, this pragmatic urgency was the deciding factor. The tech lead role, after all, is a leadership position, even when it's not a management position.

I've also seen tech leads flounder. One particularly memorable struggle was a person who was an amazing engineer, wrote great code, but hated talking to people and often got distracted by technical details. I watched him go down rabbit hole after rabbit hole, and in the meantime, the product manager took advantage of his absence to railroad the rest of the team into committing to feature delivery that was both poorly designed and way too aggressive. The project was a mess, and what did the tech lead do? He went chasing after the next refactoring, because he was sure that the problems were entirely in the way the code was structured. You probably recognize that story, because it happens everywhere. The idea that the tech lead role should automatically be given to the most experienced engineer, the one who can handle the most complex features or who writes the best code, is a common misconception that even experienced managers fall for. Tech lead is not the job for the person who wants the freedom to focus deeply on the details of her own code. A tech lead who does this is not doing her job. But what is the job of tech lead, really? What do we expect from this person?

As with many titles in software engineering, "tech lead" lacks a common definition. The best I can do is draw from my own experience and the experience of others. My job as tech lead was to continue to write code, but with the added responsibilities of representing the group to management, vetting our plans for feature delivery, and dealing with a lot of the details of the project management process. I could be the tech lead, despite not being the most senior person, because I was willing and able to take on the responsibilities of the role, while the rest of my team were more interested in staying purely focused on the software they were writing. When my team at Rent the Runway created our engineering career ladder, we consciously chose to define the role of tech lead as a set of characteristics an engineer could take on at many points on the ladder, rather than a specific level. We took this tack because we wanted to recognize that, as teams change and evolve, the tech lead role may be held by many different stages of engineer, and may be passed from one engineer to another without either person necessarily changing his functional job level. The tech lead may not have exactly the same role from company to company, or even from team to team within a company, but we know from the title that it is expected to be both a technical position and a leadership role, and that it is often a temporary set of responsibilities rather than a permanent title. So, with all that said: what *is* a tech lead? Here is the description we created at Rent the Runway:

> The tech lead role is not a point on the ladder, but a set of responsibilities that any engineer may take on once they reach the senior level. This role may or may not include people management, but if it does, the tech lead is expected to manage these team members to the high management standards of RTR tech. These standards include:
>
> - Regular (weekly) 1-1 touchbases
> - Regular feedback on career growth, progression towards goals, areas for improvement, and praise as warranted
> - Working with reports to identify areas for learning and helping them grow in these areas via project work, external learning, or additional mentoring
>
> If a tech lead is not managing directly, they are still expected to provide mentorship and guidance to the other members of the team.

The tech lead is learning how to be a strong technical project manager, and as such, they are scaling themselves by delegating work effectively without micromanaging. They focus on the whole team's productivity and strive to increase the impact of the team's work product. They are empowered to make independent decisions for the team and are learning how to handle difficult management and leadership situations. They are also learning how to partner effectively with product, analytics, and other areas of the business.

It is not required that an engineer work as a tech lead to progress, but it is the most common way for engineers to grow from senior engineer 1 -> 2 and is required to grow from senior engineer 2 to engineering lead. Realistically it is very hard to grow past senior engineer 2 without ever having acted as a tech lead, even on the individual contributor track, due to the importance at senior levels of leadership and responsibility.

Perhaps a better shorthand for this is the description used by Patrick Kua in his book, *Talking with Tech Leads* (*https://leanpub.com/talking-with-tech-leads*):

A leader, responsible for a (software) development team, who spends at least 30 percent of their time writing code with the team.

Tech leads are in the position to act as strong technical project leaders, and to use their expertise at a larger scale so that their whole team gets better. They can make independent decisions, and they play a big role in coordinating with other nonengineering partners that their team might have. You'll note that there's nothing here about specifically technical work. This is a senior engineering role, but it's a mistake to tie the notion of tech lead to one that boils down to the best or most experienced engineer on the team. You can't lead without engaging other people, and people skills are what we're asking the new tech lead to stretch, much more than pure technical expertise. However, tech leads will be working on one major new technical skill: project management. The work of breaking down a project has a lot of similarity to the work of designing systems, and learning this skill is valuable even for engineers who don't want to manage people.

If you've found yourself in the tech lead position, congratulations! Someone thinks you have what it takes to be the point person for a team. Now it's time to learn some new skills!

Being a Tech Lead

Being a tech lead is an exercise in influencing without authority. As the tech lead I am leading a team, but we all report to the same engineering manager. So not only do I have to influence my peers, but I also have to influence up to my manager to ensure we are prioritizing the right work. In a recent role this was particularly challenging because one of the first projects I tackled after becoming the tech lead was to stop all feature development and focus on technical debt. It was clear to me that the "technical debt can" had been kicked down the road for far too long; deploying new code was difficult, operating the existing services was expensive, and the on-call rotation was hellish. I believed that we needed to go slow in order to go fast in the future. However, this was not an easy sell to the other developers, who wanted to write fun new features, or to my manager, who had a constant stream of requests from our customers. I sold the idea by focusing on the different impact this would have on individual team members. For some team members it was about having a more reliable service, for others it was about iteration speed, and for others it was about reducing the on-call burden so that they could sleep through the night. When talking with my manager I emphasized the reduced operational overhead, which meant we could accomplish more feature work as a team in the future.

Becoming a tech lead required me to change my focus. Work is now less about me and working on the most technically challenging idea or the most fun project; instead, my focus is more on my team. How do I empower them? How do I remove the obstacles slowing them down? Working on a rewrite, or some new exciting feature that helped me express the full extent of my technical prowess, might have been more fun, but what the team needed at the time was to tackle technical debt and to focus on operations. In the end the initiative was incredibly successful. The team reduced the number of critical paging alerts by 50%, and in the following quarter we almost doubled the number of deploys we were able to do.

—*Caitie McCaffrey*

All Great Tech Leads Know This One Weird Trick

You're a tech lead, which means you know something about software, and your manager thinks you're mature enough to be given greater responsibility for projects. Having the technical chops and maturity is nothing, however, if you can't figure out the biggest trick of being a good tech lead: the willingness to step away from the code and figure out how to balance your technical commitments with the work the whole team needs. You have to stop relying entirely on your *old* skills and start to learn some *new* skills. You're going to learn the art of balance.

From now on, wherever you go in your career, balancing is likely to be one of your core challenges. If you want to have autonomy over your work, if you want the freedom to make choices about what you work on when, you must gain mastery over your time and how you use it. What's worse, you'll often need to balance doing things you know how to do and enjoy doing, such as writing code, with things you don't know how to do. It's natural for humans to prefer activities they've mastered, so when you have to spend less time on your current talents in favor of learning new things, it'll feel quite uncomfortable.

It can be hard to balance the work of project management and oversight with hands-on technical delivery. Some days you're on a maker's schedule, and some days you're on a manager's schedule. Through trial and error, you'll need to learn how to manage your time to provide yourself with appropriately sized blocks to work in. The worst scheduling mistake is allowing yourself to get pulled randomly into meetings. It is very difficult to get into the groove of writing code if you're interrupted every hour by a meeting.

Even with careful scheduling, you won't often have the time to focus for several-day stretches on coding problems. Hopefully, you've learned some tricks before now to help you break down your own work so that you don't need to spend multiple days of focused effort to finish technical tasks. You also know that it's important to get your team into a schedule that allows them to be focused on development for long stretches of time, because *they* will need to focus for several days on coding problems. Part of your leadership is helping the other stakeholders, such as your boss and the product manager, respect the team's focus and set up meeting calendars that are not overwhelming for individual contributors.

Being a Tech Lead 101

Let's say you're partnering with a product manager and a team of four other engineers on a big multiweek effort to launch a new initiative. The tech lead has a number of responsibilities in this scenario, depending on where you are in the project lifecycle. Sure, you'll need to write some code and make some technical decisions. But that's only one of the roles you'll play, and it's likely not even the most important one.

THE MAIN ROLES OF A TECH LEAD

Your highest priority as a tech lead is taking a wide view of the work so that you keep the project moving. How do you go from organizing and planning the code you need to write on your own to organizing and leading the overall project?

Systems architect and business analyst

In the systems architect and business analyst roles, you identify the critical systems that need to change and the critical features that need to be built in order to deliver upcoming projects. The goal here is to provide some structure for basing estimates and ordering work. You need not perfectly identify every single element of a project, but there's a lot of value in spending time thinking through the externalities and issues related to a project. This role requires you to *have a good sense of the overall architecture of your systems and a solid understanding of how to design complex software.* It probably also requires you to be able to *understand business requirements and translate them into software.*

Project planner

Project planners break work down into rough deliverables. With this hat on, you're learning to find efficient ways of breaking down the work so that the team can work quickly. Part of the challenge here is getting as much productive work done in parallel as possible. This can be tough because you are probably used to thinking about only your own work, instead of the work of groups of people. Finding places to apply agreed-upon abstractions to enable parallel work is key. For example, if you have a frontend that consumes JSON objects from an API, there should be no need for the API to be completely finished for the frontend development to begin. Instead, agree upon the JSON format and start to code to that format using dummy objects. If you are lucky, you've seen this happen before and are simply pattern-matching your previous work. At this stage, you will want to *gather input from the experts on your team,* and talk to the people who know the affected parts of the software deeply, so that they can help with the details here. You will also want to *start identifying priorities* as part of this process. Which pieces are critical, and which are optional? How can you work on the critical items early in the project?

Software developer and team leader

Software developers and team leaders write code, communicate challenges, and delegate. As projects move forward, unexpected obstacles arise. Sometimes tech leads are tempted to go to heroics and push through these obstacles themselves, working excessive overtime to get it all done. In your position as tech lead, you should *continue writing code, but not too much*. Even if you are tempted to pull a rabbit out of the hat yourself, you must communicate this obstacle first. Your product manager should know as early as possible about any possible challenges. Enlist the help of your engineering manager as needed. In a healthy organization, there is no shame or harm in raising issues early. Teams often fail because they overworked themselves on a feature that their product manager would have been willing to compromise on. As a large project nears its delivery date, there will be compromises on functionality. Start looking for opportunities to *delegate work*, especially if there is part of the system you expected to build yourself that you have not had the time to tackle.

As you can see from these descriptions, in the process of being a tech lead, you have to act as a software developer, a systems architect, a business analyst, and a team leader who knows when to do something single-handedly, and when to delegate the work to others. Fortunately you don't have to do all of these tasks at once. It may be uncomfortable at first, but you'll find a balance with time and practice.

Ask the CTO: I Hate Being a Tech Lead!

I thought becoming a tech lead would be awesome, but now my manager expects me to chase down all these details about project status, and tell her when things are going to be done, and I really hate it. Why did no one tell me that the tech lead position was so terrible?

All this new responsibility is hard, I know. I like to call this particular problem the "Stone of Triumph." (*Simpsons* fans will get my joke.) The Stone of Triumph is a metaphor for achieving recognition only to discover that recognition comes with a heavy price. While this is true at many stages of an engineering leadership career, the tech lead stage is surely one of the heaviest stones. Very rarely is the tech lead given an increase in salary or a title bump, and first-time tech leads often have no idea how hard the new responsibilities can be. As I mentioned in the definition of the position, many companies consider this to be more of a temporary title, a set of responsibilities you may take and shed several times in your career. It can be a stepping-stone necessary for promotion to more senior levels, but it is not usually a milestone that comes with immediate, tangible rewards.

Why is the tech lead role such a heavy burden? The tech lead has a much wider scope of responsibility than the senior engineer in an individual contributor position. The tech lead is called on to help architect a project, and then to go through the steps of actually planning out the work. The tech lead is expected to make sure the team fully understands the project requirements, the work is planned, and the team is effective and performing well, all without necessarily having any management responsibilities and usually without any specific training. And, realistically, most managers will expect their tech leads to continue to write almost as much code as they did before they took on the lead role. It's generally a pure increase in responsibility and scope of work. If you're a first-time tech lead, you have your hands very full.

So, congratulations, they've given you the Stone of Triumph! Fortunately, carrying around that burden will eventually make you stronger and give you skills you need to move forward in your career. It won't always be as heavy as it seems right now.

Managing Projects

I remember my very first experience with complex project management quite vividly. I was a first-time tech lead and my team was undertaking a very complex task. We had an existing system that we had scaled to its breaking point. After throwing just about every hack in the book at it, we decided it was time to figure out how to run it across several machines. This was back in the very early days of distributed systems, when most software developers really didn't know all that much about the best practices of creating them. But we had a great team of smart people, and we felt confident that we could figure this out.

We did figure it out, slowly but surely. We spent a long time thinking about design and the different ways of breaking up our computations so they made sense when computed across multiple machines. And then, one day, my boss Mike pulled me into his office and told me I needed to make a project plan.

It was one of the worst experiences ever.

I had to take this incredibly complicated set of tasks and try to figure out which ones depended on other ones. I had to think of all kinds of dependencies. How would we make it work in the complex testing framework we depended on? How would we deploy it? When did we need to order hardware to test it? How long would integration testing take? The questions just kept coming. I would go into Mike's office, sit across from him at this big wooden desk, and we would go over task descriptions, dates, and breakdowns. He would help me do some of it, and then send me off with the parts that needed more work.

This was not something I enjoyed doing. It is burned into my memory as a series of frustrating and tedious steps where I had to overcome uncertainty and the fear of making mistakes, the fear of missing pieces, in order to create a plan that would pass Mike's judgment. Then we had another round of tedious work to put it into a format that we could present to the leadership team, so that they would accept it. It almost killed me. And it was one of the most important learning experiences of my career.

Doesn't agile software development get rid of the need for project management? No. Agile software development is a great way to think about work because it forces you to focus on breaking tasks down into smaller chunks, planning those smaller chunks out, and delivering value incrementally instead of all at once. None of this means that you don't need to understand how to do project management. You'll have projects that for whatever reason can't be completed in a single sprint, or even two small sprints. You'll need to estimate project length for your management team, and give some detail on why you believe things will take that long. There are some projects, usually described by words like *infrastructure*, *platform*, or *system*, that require architecture or significant advanced planning. When faced with this kind of project, which includes many unknowns and relatively hard deadlines, you will find it doesn't fit so well into the standard agile process.

As you move forward in your career, you need to understand how to break down work that has complexity beyond the scope of what you can do as an individual. Project management for a long-running, team-based project is not what most people consider fun. I find it tedious and sometimes kind of scary. I want to be building and getting value, not trying to think about how to break down a project that still has very fuzzy implementation details. I'm afraid that I will be held accountable and that I could miss something important in the process that will make the project fail. But the alternative is the project failing slower, not faster.

Project management isn't something that needs to happen in detail for every single effort, and it's overused in some organizations. I don't even like hiring project managers because they often act as a crutch for engineers to use instead of learning to think through their future work and ask real questions about what they're doing and why, and their presence means that you have more waterfall-style projects instead of an agile process. Still, project management has to happen, and as tech lead, you should be doing it when it is needed, especially for deeply technical projects.

Ultimately, the value of planning isn't that you execute the plan perfectly, that you catch every detail beforehand, or that you predict the future; it's that you enforce the self-discipline to think about the project in some depth before diving in and seeing what happens. A degree of forethought, in places where you can reasonably make predictions and plans, is the goal. The plan itself, however accurate it turns out, is less important than spending time on the act of planning.

Back to my first project management experience. Did the project run perfectly according to plan? Of course not. There were bumps, bugs, unexpected delays, and things that we missed. However, amazingly, we still delivered the project fairly close to on time, and there was no string of sleepless nights required to get there. We managed to make the changes needed to move this complex system into a distributed deployable artifact, all while working against the master code branch with 40 other developers making concurrent changes of their own. All of this was possible because we had a great team, and we had a plan. We had thought through what success looked like, and we had identified some of the risks that might cause failure.

Since that frustrating series of meetings with Mike, I've had my own series of project planning meetings where I was the one sitting in Mike's place, and across from me was Carlo, or Alicia, or Tim. They each felt the frustration of the plan lacking detail, and they each went away and did the uncomfortable work of thinking about things that aren't code, that couldn't be perfectly predicted. And they've each led complex projects to successful outcomes thanks to this work, and are better equipped to build bigger systems and lead larger teams now that they understand what breaking down a project really means.

Take the Time to Explain

One of the last steps in a doctoral program is the defense. This is where you, the doctoral candidate, after years of research, are presenting the results of your work in front of a panel of experts in your field who will judge if the merits of your results are worthy of a PhD. Years ago, I was fortunate to receive a PhD in mathematics from one of the most prestigious Applied Mathematics programs in the Unites States. One of the judges on my panel was a renowned mathematician in the field of numerical analysis. Something he said to me after my (successful) defense has stuck with me throughout my working career (not in mathematics!). He said, "Your thesis was one of the most lucid and clear theses I've read in many years. Thank you!" I was certainly gratified but also very surprised by his words. I had assumed that he, being a world-class mathematician, would "know all about it," and just "watch" how my thesis would turn out. In fact, as he explained, he was able to do that, but only because I had taken the trouble to explain the basic ideas of the problem space and the motivations behind my ideas. I have never forgotten this lesson. Since then, after many years working in software and in large organizations, I have come to appreciate those comments even more.

We think our management "gets" what we do as technologists. Just "read the code, man!" The software we live and breathe every day ought to seem obvious to anyone working in technology, right? But it is not. Technology managers hire the best people (hopefully), who solve very difficult problems. But they don't "get" it all. I've always been surprised how grateful senior technical managers have been when I can explain some very basic modern ideas (e.g., what's this NoSQL stuff all about, and why should I care?) to them in a nonthreatening and noncondescending way.

Recently, a senior business manager at work asked me privately about why it was important for us to migrate our traditional deployed fat-client architecture to a hosted platform. He was under a lot of internal pressure to fund this effort, but he hadn't a clue why this was necessary. He was also probably too embarrassed to ask publicly. I spent two very fruitful hours explaining (without PowerPoint!). I never hesitate nowadays to take the opportunity to explain basics and motivation to senior or junior members. It educates them without making them feel small, they learn to trust my judgment and advice, and we bring about change. Taking the time to explain is very important.

—Michael Marçal

Managing a Project

Project management is the act of breaking a complex end goal down into smaller pieces, putting those pieces in roughly the most effective order they should be done, identifying which pieces can be done in parallel and which must be done in sequence, and attempting to tease out the unknowns of the project that may cause it to slow down or fail completely. You are addressing uncertainty, trying to find the unknowns, and recognizing that you are going to make mistakes in the process and miss some unknowns despite your best efforts. Here are some guidelines:

1. **Break down the work.** Get out a spreadsheet, or a Gantt chart, or whatever works for you, and start breaking down your big deliverable (say, rewriting your billing system) into tasks. Start with the biggest pieces, then break the big pieces down into smaller pieces, then break those down into even smaller pieces. You don't actually have to do it all yourself. If there are parts of the system you don't understand well, ask for help from the person who does. Get the big stuff broken down some, and then turn your attention to the ordering of the work. What can start immediately? Hand those tasks off to the people who can actually turn them into ticket-sized work.

2. **Push through the details and the unknowns.** The trick of project management is not to stop when you feel a little bit stuck, or tired of it. It *is* tiring and tedious, as I said earlier. And it's not something you probably know how to do well. So keep pushing through it past those points of irritation, boredom, and pain. A good manager will sit with you and tell you where it isn't good enough, ask questions to prompt you, or even work through some of it with you. We don't enjoy it either, but it is part of the teaching exercise. Work through the unknowns until you really feel that there is no more value to be gained in spending time on them.

3. **Run the project and adjust the plan as you go.** The value of a good planning process is that it helps you know approximately how far the project

has come, and approximately how far it is from completion. As things slip (and they always do), keep everyone apprised of the status. But now, instead of guessing how far you have to go, you can clearly point to the milestones that have been hit and outline the expected remaining work.

4. **Use the insights gained in the planning process to manage requirements changes.** You learned a lot by breaking down the project given the original set of requirements. If requirements start to change midflight, take those insights and apply them to the changes. If the changes add significant risk to the project, necessitate a bunch of new planning, or simply require a lot of additional work, be clear about the cost of those changes. If you're working toward a hard deadline, knowing roughly the effort required will help you prioritize, cut, and simplify work to get the best compromise of features, quality, and delivery date.

5. **Revisit the details as you get close to completion.** Toward the end of the project, the tedium returns. It is time to really attend to the finishing details. What is missing? What testing? What verification? Run a *premortem*, an exercise where you go through all the things that could fail on the launch of this big project. Decide where the line for "good enough" is, socialize it, and commit to it. Cut the work that falls below the "good enough" line, and focus the team on the most important final details. Make a launch plan; make a rollback plan. And at the end of it, don't forget to celebrate!

Ask the CTO: I'm Not Sure I Want to Be a Tech Lead

My manager keeps pushing me to consider being a tech lead. She wants me to run a big project. I know if I took the role I would have much less time to write code because I'd have to be in a lot of meetings and dealing with a bunch of coordination. I don't think I want this, but how do I decide?

I have a strong opinion on pushing people into management roles, which is that you shouldn't do it. If you're not ready to take on management-type responsibilities, don't take them on. There's nothing wrong with staying deep in the technology, especially if you feel like you still have a lot to learn before you become an expert.

Good managers are looking out for talented people who could be given bigger leadership roles, but sometimes this leads them to push people away from coding before they're ready. This practice can have a very negative impact on your career, because at more senior levels people who are considered "not technical enough" can find it hard to be promoted into management positions with more responsibility. It's much easier to stay in a focused individual contributor role and learn what you need to learn there than it is to try to learn all of those skills while also learning management skills.

At some point, to progress in your career, you'll probably need to do the tech lead job, even if you're interested in staying on the individual contributor (nonmanagement) career path. That doesn't mean you need to do it now. If you feel like there's plenty of purely technical learning for you to do on your team, and you'd rather work individually on this project with someone else running it, don't take the tech lead role. If, on the other hand, you don't think the individual work would challenge you technically, perhaps it's time to push yourself into learning some new skills—and the skills of the tech lead are good ones to try out.

Decision Point: Stay on the Technical Track or Become a Manager

The decision of whether to be a manager or stay on the technical track is a tough one. It's incredibly context-specific, and I can't possibly tell you what to do. However, as someone who has dreamed and lived both of these tracks, I can tell you how I imagined the roles versus what I ended up experiencing and observing. So, with the caveats that these are simply caricatures and not set in stone, let me tell you how imagination and reality diverged for me.

IMAGINED LIFE OF A SENIOR INDIVIDUAL CONTRIBUTOR

Your days are spent in a mix of deep thinking, solving hard problems that challenge you intellectually but are still fun and novel, and collaborating with other deep thinkers. It's software, so you know there will be some yak shaving, but you get to do some of the most interesting work, and you have a lot of power to choose what you work on. You love writing code, fixing code, making code go faster, and making computers do new things, and you get to spend most of your time doing that.

Because of your seniority, the managers ask you for your advice on how to approach development before it begins, so you know everything that's going on but you don't really need to deal with the details of the people building it. You're invited to just the right set of meetings where the important decisions are made, but not so many as to disrupt your flow. The more junior developers look up to you and hang on your every word, taking your feedback but not imposing too much on your deep thinking time.

Your upward trajectory is never slowed, and there are always new big problems that you can solve to show off your value to the organization. You work hard, but are rarely called upon to stay late or work weekends, because as we all know it is impossible to do quality, thoughtful work for too many hours a week. When you do work late, it is because you are just so caught up in the flow that you can't wait to finish the feature at hand or fix the bug you just found.

You get to write books, give talks, and create open source work—and with some luck and persistence, you earn a bit of industry-wide fame. No one cares that you're a bit awkward or shy or expects you to evolve your communication style much, because what you say is so important. Everyone in your organization knows who you are, understands how valuable your work is, and is deferential to your opinions.

In short, you have the perfect balance of engaging work, fame, and accumulated expertise that makes you invaluable and respected, highly paid, and influential.

REAL LIFE OF A SENIOR INDIVIDUAL CONTRIBUTOR

When you find the right project, and the right lifecycle of the right project, your life is great. You are challenged, and you get to learn new things. You have a lot of control over your day-to-day, and certainly fewer meetings than your management counterparts, but your days are not always spent in a blissful flow state. For every project there's the period where you have the idea and you're selling it to people, trying to convince them that it is the right approach. Or you've implemented the system, but now you need to get other teams to start using it, so you sit with them for days showing them the ins and outs, explaining why it is useful, and trying to convince them to lobby their manager for time to adopt it.

Your upward trajectory is not as fast and easy as you had hoped it would be. In fact, it is pretty slow. Those big projects that prove you to be an invaluable architect are hard to find. The team doesn't need a new programming language, a new database, or a new web framework. Your manager isn't great at handing you plum assignments that showcase your talents to the whole organization; she expects *you* to tell *her* where these opportunities live. Discovering good projects seems to be a matter of luck. Pick the wrong project, and you spend months or even years on something that might get cancelled despite all of your best efforts.

You are a little jealous of your friends in management who seem to be getting promoted faster while they keep growing their teams.

The other developers are a mixed bag. You're a nice person, so some of them admire you and listen to your opinions, but others seem to be jealous of your influence. New developers either want too much of your time or seem to be scared of you for whatever reason. There's definitely some competitiveness with your peers around who gets to lead big, interesting projects.

Your manager is kind of a pain. She isn't terribly supportive of your desire to open source a system because you think it provides a new twist on logging that the industry needs, and she suggests that if you want to give talks or write books perhaps you need to spend some of your personal time on those efforts. She seeks out your feedback on technical matters, but sometimes forgets to tell you about new initiatives until it's too late for you to put in your two cents. You suspect that you are missing out on crucial information because you aren't in the right meetings, but every time she invites you to sit in those meetings you remember how boring and inefficient they are, and how much valuable focus time you're losing. And she doesn't have much patience for your desire to be free of tedious work like answering email, interviewing, or responding to code reviews promptly.

Still, you get to build things most of the time. You get to spend your time focused on technical problems, systems design, and engineering issues, and you don't have to do all that much dealing with people or sitting in boring meetings. You can often choose your projects and can easily move between teams if you want something new. And you just found out that you get paid more than your manager! So, life isn't all bad.

IMAGINED LIFE OF A MANAGER

You have a team, you have control, you can make the decisions, and you can finally get others to do things your way. Your team respects you and is happy to yield to your authority in all matters. You think they should write more tests? You tell them, "Write more tests," and they do it! You want to make sure that everyone is treated fairly despite their gender, race, and so on? You make sure that it happens, and fire anyone who crosses the line and creates an environment that is unhealthy for the rest of the team.

Because you care about people, they know that you're always trying to do your best for them even when they disagree with you. They give you the benefit of the doubt, and come to your 1-1s with open feedback for you when you're screwing up, and eager to receive feedback from you. Dealing with people is stressful, sure, but they know you care about them, so it's also highly rewarding. You see the impact of your coaching happen quickly now that you are in this position of authority.

When you see another manager doing something that seems wrong, you are free to go give him advice in the same way you would talk to another engineer who needed help on a system design. Other managers are always interested in hearing what you think, and they can see how effectively you've gotten your team working, how clearly you care about the health of the organization, and how genuine your interest is in just making everyone better.

Your manager gives you plenty of coaching but rarely steps in to tell you what to do. The minute that you feel ready to take on a bigger team, your manager is open to giving you more people and expanding your organization. She hands you goals that are clear and rarely changes things. Even though you have a lot of responsibilities, you still have some time to write blog posts and give talks, and this is encouraged because it will help your team hire and improve your standing in the tech industry.

In short, you get to make decisions, you create the culture, and your effectiveness is evident to all around you, making your path to promotion quick and your career exciting and lucrative.

REAL LIFE OF A MANAGER

You have a team. You have some control, but you've quickly discovered that getting people to do something is harder than just telling them to do it. You seem to have given up all control over your own day-to-day. Mostly you spend all day in meetings. You knew this was coming, but it's not until you live it that you really understand what it means. When you only had a small team you were able to balance things and still write code, but as your team has grown you've lost touch with the code. It gnaws at you as something you should be doing, but there's no time. Every time you snatch a few hours to write code, you realize that it would be irresponsible to check it in and make the team support it, so at best you snatch a script here, debug an issue there. Having the focus to build something big yourself is a distant memory.

You can make decisions—well, some decisions. Realistically, you can maybe narrow down the things that will get decided. You can focus your team on some things, like writing better tests, but they still have a product roadmap to implement, and they have their own ideas for what technical tasks should be prioritized. So, more than making decisions yourself, you're helping the team make decisions. Your manager gives you goals but then sometimes changes those goals completely, and it's up to you to explain the changes to the team.

You do set the standards for culture on your team, which is good and bad. It's good when they take after your best aspects, and it's bad when you realize that your team is also mirroring your faults.

Your team does not naturally just agree with you, respect you, or even like you. You realize that authority requires more than a title. You find yourself scrambling to motivate them through tough periods when the projects aren't going well, or when you have to tell individuals that they aren't ready to be promoted just yet, that they aren't getting a raise, that there's no bonus this year. Some of them don't bother to tell you when they're unhappy; they just get fed up and quit before you've noticed there's anything wrong. When the company is doing well, and you have lots of money to pay, and there are plenty of exciting projects, life is great; but when things are stressful, you see how little power you have to make people happy. And what's worse, you can't even fire people without going through a crazy HR process! Still, you can see that your work matters to some of them, that they are happier and more successful because of your coaching. These little wins sustain you through the tough times.

Other managers are not interested in your feedback. In fact, they find you meddling and get competitive when they think you're encroaching on their turf. Your own manager does not agree that you're ready for a bigger team, and can't really explain why; his coaching skills leave a lot to be desired. Maybe he is just worried that you will outshine him? He definitely does not want you spending all of your time giving talks, though—he gets annoyed when you are out of the office too much, whatever the value the team might get from it. The politics of figuring out how to lead without undermining your peers or your boss are trickier than you expected. But if you can get that bigger team, you know you will get that promotion, so at least your path is clear. When you discovered that the staff engineer who works for you makes more than you do, you almost lost it, so you'd better figure out how to get that bigger team fast. Otherwise, what is the point of all of this stress and nonsense?

My final advice is to remember that you can switch tracks if you want. It is common for people to try out management at some point, realize they don't enjoy it, and go back to the technical track. Nothing about this choice has to be permanent, but go in with your eyes wide open. Each role has benefits and drawbacks, and it's up to you to feel out what you enjoy the most.

Good Manager, Bad Manager: The Process Czar

The process czar believes that there is one true process that, if implemented correctly and followed as designed, will solve all of the team's biggest problems. Process czars may be obsessed with agile, Kanban, scrum, lean, or even waterfall methods. They may have a very precise idea of how on-call should work, how code reviews must be done, or how the release process has to operate. They tend to be very organized and comfortable with details, and they're good at knowing the rules and following them precisely.

Process czars are often found in QA, helpdesk, or product management groups. They're also common in consulting agencies and other places where measurement of specific work progress is highly rewarded. They may be operationally focused, although in my experience, there are relatively few of these folks inside of your classic systems operations teams. They can be incredibly valuable members of a project management team because they tend to make sure that no task is forgotten and that everything is wrapped up in the way it should be.

Process czars struggle when they fail to realize that most people are not as good at following processes as they are. They tend to blame all problems on a failure to follow the best process, instead of acknowledging the need for flexibility and the inevitability of unexpected changes. They often get focused on easy-to-measure things, such as hours in the office, and miss the nuances that they fail to capture when focusing on the things that are easy to measure.

Engineers who believe in the "right tool for the job" sometimes turn into process czars when they become tech leads, seeking out the right tool to solve all issues with planning, focus, time management, and prioritization. They try to stop all work while they search for the perfect process, or constantly push new tools and processes on the team as solutions to the messier problems of human interactions.

The opposite of the process czar is not a manager who gives up on process completely, but rather someone who understands that processes must meet the needs of the team and the work. Ironically, while "agile" is often implemented in a rigid way, the principles of the Agile Manifesto (*http://agilemanifesto.org/*) are a great summary of healthy process leadership:

- Individuals and interactions over processes and tools

- Working software over comprehensive documentation

- Customer collaboration over contract negotiation
- Responding to change over following a plan

As a new tech lead, be careful of relying on process to solve problems that are a result of communication or leadership gaps on your team. Sometimes a change in process is helpful, but it's rarely a silver bullet, and no two great teams ever look exactly alike in process, tools, or work style. My other piece of advice is to look for self-regulating processes. If you find yourself playing the role of taskmaster—criticizing people who break the rules or don't follow the process—see if the process itself can be changed to be easier to follow. It's a waste of your time to play rules cop, and automation can often make the rules more obvious.

As the manager of a process czar, help that person get more comfortable with ambiguity. As with many of the manager pitfalls, an obsession with process can be related to a fear of failure and a desire to control things to prevent the unexpected. If you are honest and make it clear that it's safe to fail and to be imperfect, that's often enough to get your process czar to relax a little bit and let some ambiguity in. It is very important to keep process czars from spending all of their time seeking out the perfect tool or process, and especially important to make sure that they aren't punishing their teams for failing to follow processes.

How to Be a Great Tech Lead

Great tech leads have a number of characteristics, but these are the most important.

UNDERSTAND THE ARCHITECTURE

If you go into a tech lead role and you don't feel that you fully understand the architecture you are supporting, take the time to understand it. Learn it. Get a sense for it. Visualize it. Understand its connections, where the data lives, how it flows between systems. Understand how it reflects the products it is supporting, where the core logic for those products lives. It's almost impossible to lead projects well when you don't understand the architecture you're changing.

BE A TEAM PLAYER

If you're doing all of the interesting work yourself, stop. Look at the tricky, boring, or annoying areas of technical need and see if you can unstick those areas. Working on the less exciting parts of the code base can teach you a lot about where the process is broken. With boring or frustrating projects, there's often something obvious that can be spotted and fixed if an experienced person takes the time to look at them. If you're only doing the most boring work, stop that, too. You're a senior engineer who has a lot of talent as a developer, and it's reasonable for you to take on some of the harder tasks. You want to encourage others on your team to learn the entire system, and you want to give them chances to stretch themselves, but you needn't always be self-sacrificing in what you choose to work on. Give yourself a fun task occasionally, as long as you know you have the time to do it well.

LEAD TECHNICAL DECISIONS

You'll be involved in most major technical decisions for your team. Involved, however, is not the same thing as being the person who makes all of them alone. If you start to make all of the technical decisions without soliciting the input of your team, they'll resent you and blame you when things go wrong. On the other hand, if you make no technical decisions and leave everything up to the team, decisions that could have been made quickly can drag on without resolution.

Determine which decisions must be made by you, which decisions should be delegated to others with more expertise, and which decisions require the whole team to resolve. In all of these cases, make it clear what the matter under discussion is, and communicate the outcome.

COMMUNICATE

Your productivity is now less important than the productivity of the whole team. Often, this means that you pay the price of communication overhead. Instead of having every team member sit in a meeting, you represent the team, communicate their needs, and bring information from that meeting back to the team. If one universal talent separates successful leaders from the pack, it's communication skills. Successful leaders write well, they read carefully, and they can get up in front of a group and speak. They pay attention in meetings and are constantly testing the limits of their knowledge and the knowledge of the team. Now is a great time to practice your writing and speaking skills. Write design documents and get feedback on them from better writers. Write blog posts for your tech blog or your personal blog. Speak in team meetings, speak at meetups, and get practice standing up in front of an audience.

Don't forget to listen during all of this communication. Give others a chance to speak, and hear what they say. Practice repeating things back to people to ensure you understand them. Learn how to hear what someone says and rephrase it in your own words. If you aren't a good note taker, you may need to become one. It doesn't matter whether you choose to dive deep into technology, or become a manager—if you can't communicate and listen to what other people are saying, your career growth from this point on will suffer.

Assessing Your Own Experience

- Does your organization have tech leads? Is there is a written job description for this role? If so, what does it say? If not, how would you define the role in your organization? How would a tech lead define the role?

- If you are considering becoming a tech lead, are you ready to push yourself? Are you comfortable spending some of your time outside of the code? Do you feel like enough of an expert in your code base to successfully lead others as they work in it?

- Have you asked your manager what he or she expects from the tech lead?

- Who is the best tech lead you ever worked with? What are some things that person did that made him or her great?

- Have you worked with a frustrating tech lead? What did he or she do that frustrated you?

4

Managing People

New engineering managers think of the job as a promotion, giving them seniority on engineering tasks and questions. This is a great approach for ensuring they remain junior managers, and unsuccessful leaders at that. It's hard to accept that "new manager" is an entry-level job with no seniority on any front, but that's the best mindset with which to start leading.

—MARC HEDLUND

Congratulations! You've progressed to the level where people trust you to manage other humans. Maybe you've had some training from your HR department about some basics of management. Maybe you've had some great managers in the past you want to emulate. But now the rubber hits the road, and it's time to put all those thoughts and ideas into action.

First, let's focus on managing individuals. There are a bunch of books out there that will give you more thoughts and ideas about this topic; my goal here is to give you the basic elements of management as I see them. Once you're in the management hot seat, how should you think about performing the basic tasks of managing people?

Part of your focus throughout this period of adjustment to management is to figure out your own management style. Many of you will be learning how to manage individuals while simultaneously being responsible for running a team. In the next chapter, we'll talk more about the challenges of dealing with the team as a whole, as well as how the technical side of your role might be changing, but it's important to start by considering the individuals. After all, your team is only as healthy as its individuals, and as the individual manager, you'll have a huge impact on each person.

We'll talk about the main tasks required to manage people:

- Taking on a new report

- Holding regular 1-1s

- Giving feedback on career growth, progression toward goals, areas for improvement, and praise as warranted

- Working with reports to identify areas for learning and helping them grow in these areas via project work, external learning, or additional mentoring

Starting a New Reporting Relationship Off Right

The first thing that happens when you start managing is that you take on people as direct reports. These may be people you've been working with for a while, or they may be completely unknown to you. As you go through your management career, you'll repeatedly experience having someone new start to report to you. How do you get to know this person quickly so you can manage him best?

BUILD TRUST AND RAPPORT

One strategy is to ask a series of questions that are intended to help you get to know the aspects of the person that impact your ability to manage him well. These questions might include:

- How do you like to be praised, in public or in private?

    Some people really hate to be praised in public. You want to know this.

- What is your preferred method of communication for serious feedback? Do you prefer to get such feedback in writing so you have time to digest it, or are you comfortable with less formal verbal feedback?

- Why did you decide to work here? What are you excited about?

- How do I know when you're in a bad mood or annoyed? Are there things that always put you in a bad mood that I should be aware of?

    Maybe a direct report fasts for religious reasons, which sometimes makes him cranky. Maybe he always gets stressed out while on-call. Maybe he hates reviews season.

- Are there any manager behaviors that you know you hate?

    If you asked me this question, my answer would be: skipping or rescheduling 1-1s, neglecting to give me feedback, and avoiding difficult conversations.

- Do you have any clear career goals that I should know about so I can help you achieve them?

- Any surprises since you've joined, good or bad, that I should know about? *Things like: Where are my stock options? You promised me a relocation bonus and I haven't gotten it yet. Why are we using SVN and not Git? I didn't expect to be so productive already!*

For more ideas, see Lara Hogan's excellent blog post on the topic (*http://lara hogan.me/blog/first-one-on-one-questions/*).

CREATE A 30/60/90-DAY PLAN

Another approach that many experienced managers use is to help their new reports create a 30/60/90-day plan. This can include basic goals, like getting up to speed on the code, committing a bug fix, or performing a release, and is especially valuable for new hires and people transferring from other areas of the company. The more senior the hire, the more he should participate in creating this plan. You want him to have some clear goals that will show whether he's learning the right things as he gets up to speed. These goals will also require some work from you and from the team, because it's very rare that everything is self-evident, well documented, and totally obvious to a newcomer.

Unfortunately, sometimes you will mishire a person. Having a clear set of expected goals for your new hires that you believe is achievable in the first 90 days will help you catch mishires quickly, and make it clear to you and to them that you need to correct the situation. Create a set of realistic milestones based on your prior hires, the current state of your technology and project, and the level of the person coming in.

ENCOURAGE PARTICIPATION BY UPDATING THE NEW HIRE DOCUMENTATION

For early- to mid-career hires, one aspect of onboarding will likely include contributing to the team's onboarding documentation. A best practice in many engineering teams is to create a set of onboarding documents that are edited by every new hire as he gets up to speed. He edits the documentation to reflect processes or tools that have changed since the last hire, or points that he found confusing. As the manager, you don't necessarily need to be the person walking the new hire through this process—that could be a job for a peer, mentor, or tech lead—but you may be the person to get this process in place, and you'll need to reinforce it for everyone who joins the team.

COMMUNICATE YOUR STYLE AND EXPECTATIONS

Your new hire needs to understand your expectations and your style just as much as you need to understand his. You'll each need to adjust a little bit to meet the other, but if the new hire doesn't know what you expect from him, he can't deliver what he needs to deliver. This expectation setting on your part should include specifics like how often you want to meet with him, how the two of you will share information, and when and how often you'll want to review his work. If you expect him to send you a weekly summary of his progress via email, tell him. Help him understand how long he should work alone trying to solve a problem, and at what point he should ask for help. For some teams this might be an hour, and for others it might be a week.

GET FEEDBACK FROM YOUR NEW HIRE

One final piece of advice: get as much feedback as you can about the new hire's perspective on the team in that first 90 days. This is a rare period, where a new person comes in with fresh eyes and often sees things that are hard for the established team members to see. On the other hand, remember that people in their first 90 days lack the context that the overall team possesses, so take their observations with the requisite grain of salt, and definitely don't encourage people in this period to criticize the established processes or systems in a way that makes the existing team feel attacked.

Communicating with Your Team

> *Regular 1-1s are like oil changes; if you skip them, plan to get stranded on the side of the highway at the worst possible time.*
>
> **—MARC HEDLUND**

HAVE REGULAR 1-1S

I had an interesting conversation once with a friend, another CTO who has a lot of management experience. He sheepishly admitted to me that he didn't like doing regular 1-1s because he himself had always resented being forced to do 1-1s with managers that he felt he didn't need. "Regular 1-1s are like going to a psychiatrist when you're fine and discovering you have depression." I bow to his experience. It is absolutely true that every person and team is different: they need different things, have different communication styles, and focus on different things. That being said, if you're not a CTO with years and years of management experience, you should probably start by assuming that you need to do regularly scheduled 1-1s.

SCHEDULING 1-1S

The default scheduling for 1-1s is weekly. I encourage you to start with weekly 1-1s and adjust the frequency only if both of you agree that this is more than you need. Weekly means that you talk frequently enough to keep the meetings short and focused, and it gives you room for the occasional missed week. When you meet less frequently, any missed 1-1s *must* be rescheduled, which is usually a drag on both of you.

Try to schedule your 1-1s for times when you and your report are both likely to be in the office. Mondays and Fridays are bad days for 1-1s because people tend to sometimes take long weekends and miss these days. I prefer to do 1-1s in the morning before things get busy, in order to avoid having the schedule slip or being forced to reschedule due to other things coming up. However, morning 1-1s only work with people who also get in early and those who don't have morning standup meetings to work around. Respect the "maker schedule" for your reports and try to give them 1-1 times that aren't likely to be right in the middle of their productive workflow hours.

ADJUSTING 1-1S

Like most things in life, 1-1s aren't just "set it and forget it." There are a number of factors to take into consideration:

How often do you interact with this person offhand during the week?
    If you interact with her frequently, you may not need a weekly set-aside time to chat.

How much coaching does this person need?
    A very junior person who's just joined the team might appreciate more time than a senior person who's in a groove. On the other hand, a senior person who is pushing through a difficult new project may appreciate more dedicated time for you to help her with some of the details of that work.

How much does this person push information up to you?
    A person who's not good at pushing information up may need more face time to do so.

How good is your relationship with this person?

> Be careful here. Some people assume that good relationships require very little attention, and spend all of their time on their bad relationships. But there are plenty of people, myself included, who feel a strong need for regular 1-1 time even in good relationships. Just because you think things are going smoothly with this person doesn't mean that she agrees. Don't make the fatal error of spending all your time with your problem employees and ignoring your stars.

How stable or unstable are things in the team or the company?

> One of the topics of discussion in your 1-1s will be company news. Especially in times of rapid change or uncertainty, make sure you take the time to answer any questions folks may have. Keeping your 1-1s regular through times of uncertainty will help stabilize your team and slow down the rumor mill.

Different 1-1 Styles

Now that you have these 1-1 meetings scheduled, what should you actually use them for? I've seen several distinct styles of 1-1 meetings, and the type of 1-1 that's most effective depends as much on the manager as it does on the managed.

THE TO-DO LIST MEETING

One or both parties comes in with a list of objectives to cover, and the parties cover these objectives in order of importance. Updates are given, decisions are made or discussed, planning happens. This style follows the "don't waste time with pointless meetings" mandate and ensures that things will happen. The downside, of course, is that sometimes you wonder why you needed to do this in a synchronous manner. Often the list is somewhat artificial and made up of things that could have been handled via chat or email. If you decide to adopt this style, make sure that the list you bring and the lists you encourage your reports to bring are meaningful for 1-1 discussions. Make sure there are nuances that deserve verbal communication in the 1-1 setting.

In general this style is very professional and efficient, if sometimes a bit cold. It forces your reports to think beforehand about the meeting and what they might want to discuss. One manager I knew used a shared Google spreadsheet to keep a running list of topics for discussion that both he and his reports had access to, so each could add to the list whenever a thought came up, and they would review it during the 1-1. It also gave both parties a chance to see what was on the other's mind before the 1-1 happened, so that they could prepare.

THE CATCH-UP

I am not a very organized person, and 1-1s with strict requirements for to-do lists don't resonate with me. I'm happy to adopt these requirements if my reports want them, but I prefer a more fluid style. My goal in a 1-1 is first to listen to anything my direct reports want to discuss. I want the meeting to be driven by them, and I want to give them space to bring up whatever they feel is important. I view a 1-1 session as much as a creative discussion as a planning meeting. The downfall of the rambling 1-1 is that, if it's left unchecked, it can turn into a complaining session or therapy. Empathetic leaders can sometimes allow themselves to get sucked into an unhealthy closeness with their direct reports. If you start focusing a lot of energy on hearing reports' complaints and commiserating, you're quite possibly making the problem worse. You don't have to have a to-do list, but problems in the workplace need to be either dealt with or put aside by mutual agreement. There is very little value to repeatedly focusing on drama.

THE FEEDBACK MEETING

Sometimes your 1-1s will be devoted to informal feedback and coaching. It's good to hold these kinds of meetings at a regular interval, especially for your early-career employees. Quarterly is frequent enough to give the topic attention without it feeling like all you talk about is career development. Many companies force specific individual goal-setting processes on everyone, so you can use this time to review progress toward goals, whether they are formal or personal.

If you have an employee with performance issues, feedback meetings should happen more frequently, and if you're thinking of firing someone I advise you to document these feedback meetings. That documentation will include the issues you discussed and the expectations that you set with the person, in writing and sent to the person (usually via email).

As much as possible, when someone does something that needs immediate corrective feedback (insulting a colleague, missing a critical meeting, using inappropriate language), don't wait for the 1-1 to provide that feedback. If you see or hear about a direct report doing something you want to correct, try to approach that person soon after. The longer you wait, the harder it will be for you to bring it up, and the less effective the feedback will be. The same goes for praise! When something goes well, don't save up your praise—give it freely in the moment.

THE PROGRESS REPORT

When you get to the stage where you're managing managers, a lot of your 1-1 meetings will be diving into details of projects they're overseeing that you don't have time to dig into on your own. When you are managing only a handful of individuals, the only time you should be using a 1-1 to do progress reporting is when you have someone who's off on a side project that you're not personally overseeing. Getting progress reports from people you're already working closely with is a waste of time because all you're hearing about is the delta of work between now and the last standup or project review. If your 1-1s are frequently status updates, try breaking out of the habit by asking your reports to prepare answers to questions that are unrelated to the current project status, or ask them to come prepared with questions for you to answer about the team, the company, or anything else. And for those rare people who just really don't have much to talk about except for progress, you may want to take that as a sign to meet less frequently.

GETTING TO KNOW YOU

Whatever type of 1-1 you do, leave room to get to know the person reporting to you as a human being. I'm not suggesting that you should pry into your reports' personal lives, but show them that you care about them as individuals. Let them talk about their family, friends, hobbies, pets. Get to know their career so far, and ask them about their long-term career goals. It doesn't have to just be a focus on the next skill or promotion. Show that you are invested in helping them now and in the future.

MIX IT UP

For variety, you can do your 1-1s as walking meetings, or over coffee or lunch to get out of the office. Just remember that when you're not taking notes you'll probably forget some important things, so try not to rely on these for critical conversations. Many of us have overstuffed office spaces with very few private conference rooms, but as much as possible, do your 1-1 meetings in private so that you can feel free to discuss sensitive topics without worrying about being overheard.

One final piece of advice: try to *keep notes in a shared document*, with you the manager playing note taker. For each person you manage, maintain a running shared document of notes, takeaways, and to-dos from your 1-1s. This is helpful for you to keep context about what has happened, and is useful for remembering when and what feedback was given. It will also be an essential historical record to refer to when you're writing reviews or delivering feedback. If you're distracted by having a computer open in front of you during the meetings, just leave time at the end to add notes.

Good Manager, Bad Manager: Micromanager, Delegator

Jane has given her tech lead Sanjay a big project to manage. It needs to be done by the end of the month, which should be fine, but Jane is worried that the deadline will slip. So she starts attending all of the standups that she normally doesn't go to, and asks questions directly of the team about their blockers. She looks through the project tickets and makes a bunch of comments, and even reassigns some of them to other team members. When she discovers that Sanjay and the product manager decided to deprioritize a feature, Jane decides that it's time for her to take over the project, and she tells Sanjay that from now on she will be managing the day-to-day.

It's no surprise when, despite the fact that the project shipped successfully, Sanjay tells Jane that he feels like he doesn't want to be a tech lead anymore. In fact, he seems pretty low-energy, and his normal engagement and hard work are replaced by him leaving early and saying nothing in meetings. Her best team member has become a low performer seemingly overnight. What happened?

Micromanagement creeps up on you. A high-stress project that can't be allowed to slip seems at risk, and so you step in to correct it. You delegate something, but then discover that you don't like the technical choices the team has made to implement it, so you tell them to rewrite it. You force everyone to come to you before making decisions because they just can't be trusted to do the right thing, or there have been too many mistakes and you always end up paying the price.

Now, let's take a look at Jane's colleague, Sharell. Sharell has given Beth her first big project to run. Sharell knows that this project needs to ship on time, but instead of sitting in on every meeting and tracking every detail, Sharell works with Beth to determine which meetings she should attend, and helps Beth understand which details to escalate to Sharell. With this support, Beth feels confident running the project, but also knows that Sharell has her back, and when things get stressful toward the end of it, Beth enlists Sharell's help to cut scope and get the project out on time. Beth leaves the experience feeling more confident, and ready to take on bigger projects and work harder for Sharell.

Jane's and Sharell's decisions highlight the subtle differences between the micromanager and the effective delegator. Both Jane and Sharell are attempting to delegate the management of a high-priority project in order to train a new leader on their team. However, Jane ultimately never gives up control and undermines Sanjay, while Sharell makes it clear to Beth what the goals are and what her responsibilities are, and provides support and guidance to help Beth succeed.

The hardest thing about micromanagement is that there are times when you need to do it. Junior engineers often thrive under detailed oversight because they want that specific direction. Some projects go off the rails, and you occasionally need to override decisions made by your reports that could have big negative repercussions. However, if micromanagement is your habit, if it's your default approach toward leading your team, you'll end up like poor Jane, accidentally undermining the very people you need to be growing and rewarding.

Trust and control are the main issues around micromanagement. If you're micromanaging someone, chances are you're doing it because you don't trust that a task will be done right, or you want to very tightly control the outcome so that it meets your exact standards. This happens a lot when talented engineers become managers, especially if they pride themselves on their technical skills. If your value to the team has shifted from the thing you're good at (writing code) to the thing you don't yet know how to do well (managing people), it can be tempting to treat your reports as if they should be mini-mes. When a deadline slips, as it inevitably will, you view it as a failure to control the situation precisely enough, and so ratchet up the attention. You catch something not going the way you expected, and it seems to reinforce your belief that micromanaging the team is an appropriate use of your time.

Autonomy, the ability to have control over some part of your work, is an important element of motivation. This is why micromanagers find it so difficult to retain great teams. When you strip creative and talented people of their autonomy, they lose motivation very quickly. There's nothing worse than feeling like you can't make a single decision on your own, or feeling like every single piece of work you do has to be double- and triple-checked by your manager.

On the other hand, *delegation is not the same thing as abdication*. When you're delegating responsibility, you're still expected to be involved as much as is necessary to help the project succeed. Sharell didn't just abandon Beth—she helped Beth understand the responsibilities in the new role and was there as needed to support the project.

Practical Advice for Delegating Effectively

It's important to remember that being a good leader means being good at delegating.

USE THE TEAM'S GOALS TO UNDERSTAND WHICH DETAILS YOU SHOULD DIG INTO

When you feel like you want to micromanage, ask the team how they're measuring their success and ask them to make that visible to you on an ongoing basis. Then sit on your hands if you must, but wait a week or two to see what they give you. If they have nothing to share, it's a sign that you may need to do a course correction, which probably means digging into more details.

How do you decide when to ask for this information to begin with? My philosophy is simple: if the team is making progress on its goals, the systems are stable, and the product manager is happy, I rarely dig into the details beyond a cursory overview. However, that requires goals with a plan for people to be making progress against, and a product manager who can give you another perspective. When you are managing a team that doesn't have a clear plan, use the details you'd want to monitor to help them create one. What are you holding them accountable against this month, this quarter, or this year? If you can't answer that question, the first step is to help the team create those goals.

GATHER INFORMATION FROM THE SYSTEMS BEFORE GOING TO THE PEOPLE

As engineers, we have an advantage because systems can push valuable information up without the team needing to do much of anything. If you want to know the status of work, look at the version control system and the ticketing system. If you want to know how stable the systems are, subscribe to information about the alerts, look at the metrics, follow what has happened in on-call. The worst micromanagers are those who constantly ask for information they could easily get themselves. It's OK to ask for status summaries and OK to use your team as a way of surfacing the most important information from all of these sources, but use a light touch. The team will not be productive or happy spending half their time gathering information for you that you could easily find yourself. And remember, this information is just a piece of the context, not the whole picture, and it means nothing without the goals just discussed.

ADJUST YOUR FOCUS DEPENDING ON THE STAGE OF PROJECTS

If you're managing a single team or two directly, you should know all of the details of the project status as part of your regular team processes (like morning standup meetings). Different details are important at different project stages. In the beginning and design stages of a project, you may want to be more involved in order to facilitate a good set of project goals or a good system design. When you're close to a delivery date, progress details become more important, because there are more decisions to be made and those specifics convey much more actionable information. During the normal workflow, though, it's usually enough to know what's moving forward and what is taking longer than expected, especially if you can use that information to reshuffle work or help a struggling team member.

ESTABLISH STANDARDS FOR CODE AND SYSTEMS

I'm one of those deeply technical managers, and I have opinions about the way systems should be built and operated. Letting go has been hard for me, so I developed some guidelines to help me feel better about the structure around these issues. Developing basic standards as a team helps everyone communicate with one another in code and design reviews, and it depersonalizes the process of providing technical feedback. For me, basic standards meant things like how much unit testing we expected to happen with each change (generally speaking, as some tests were always required), and at what point technical decisions should be reviewed by a larger group (like when someone wants to add a new language or framework to the stack). As with goal setting, putting standards in place here helps people know which details are important to think about when they're creating the technology.

TREAT THE OPEN SHARING OF INFORMATION, GOOD OR BAD, IN A NEUTRAL TO POSITIVE WAY

Consider this scenario: Jack is having a hard time with a project, but hasn't been asking for help with his problems. You finally hear about his struggles. At this point, it's appropriate to tell Jack that he needs to be more proactive in sharing his progress, even if it means admitting he's struggling. You could have Jack give you daily updates as a way of helping, but I would only use that much structure for a brief period. The goal here isn't to punish him with micromanagement for his failure to communicate status, because all you're doing is punishing yourself and hindering his ability to be held accountable for his own work. Instead, your goal is to teach Jack what he needs to communicate, when, and how. A word of caution, though; if you treat a struggling engineer or project as a massive failure on the part of the individual or manager, she is going to feel that blame and criticism, and instead of giving you more information in the future, she'll keep hiding it from you as a way of avoiding blame until it's too late. Hiding important information intentionally is a failure, and getting stuck on a problem or making a mistake is often just an opportunity for learning.

In the long run, if you don't figure out how to let go of details, delegate, and trust your team, you're likely to suffer personally. Even if your team doesn't quit, you'll end up working longer and longer hours as your responsibilities increase. If you're already in this situation, try limiting the hours you'll let yourself work in a week. If you were only allowed to work 45 hours this week, what would you do with those 45 hours? Would you really spend five of them nitpicking a junior developer's code? Would you pore over the details of some project that is going smoothly, searching for any minor error? Or would you direct your attention to bigger problems? Would you take some of those hours to focus on the future instead of the details of the current moment? Your time is too valuable to waste, and your team deserves a manager who is willing to trust them to do things on their own.

Creating a Culture of Continuous Feedback

When I say *performance review*, what goes through your head? Do you cringe? Do you roll your eyes about the wasted time, or groan at the thought of doing all the work? Do you get a rush of fear over what surprising new flaws you will hear about? Or do you get a little bit of nervous excitement to hear what people think about you?

If performance reviews make you shudder, you're not alone. Unfortunately, the review process is not something that every manager takes seriously or handles in a mature way. Now that you're managing people, you have a lot of power to shape the experience your direct reports have with reviews. That experience starts long before the reviews are written. It starts with continuous feedback.

Continuous feedback is, more than anything, a commitment to regularly sharing both positive and corrective feedback. Instead of saving these kinds of comments for the review cycle, managers and peers are encouraged to note when things are going well and raise issues as they happen. Some companies have started to adopt software that makes it easy for teams to provide continuous feedback and track that feedback over time, but the most important thing is that the team has adopted a culture of providing feedback frequently. For you, as a new manager, getting into the habit of continuous feedback is training you to pay attention to individuals, which in turn makes it easier to recognize and foster talent. You're also practicing the art of having small and occasionally tricky conversations with individuals about their performance. Few people are comfortable with providing one-on-one praise or correction, and this helps you get over the feeling of awkwardness.

There are some steps you can take to be great at giving continuous feedback:

1. **Know your people.** The first required part of successfully giving continuous feedback is a basic understanding of the individuals on your team. What are their goals, if any? What are their strengths and weaknesses? At what level are they currently operating, and where might they need to improve to get to the next level? You can get some of this knowledge by reading their previous performance reviews if you have them, but you'll also want to sit down with every person on your team and ask for his or her perspective on all of these questions. This understanding gives you a baseline you can use to frame your feedback, and helps you find some things that you might want to focus on.

2. **Observe your people.** You can't give feedback if you aren't paying attention. If anything, I think the best outcome of attempting a continuous feedback cycle is not necessarily the actual feedback generated, but rather that the effort forces you to start paying attention to the individuals on your team. Starting this habit early in your management career, while you still may have only a few people to manage, helps you build up those observational muscles. Practice looking for talents and achievements on your team, first and foremost. Good managers have a knack for identifying talents and helping people draw more out of their strengths. Yes, you'll also want to look for weaknesses and areas for improvement, but if you spend most of your time trying to get people to correct weaknesses, you'll end up with a style that feels more like continuous criticism.

   Sometimes it helps to have a goal, so task yourself with regularly identifying people who deserve praise. Adopting a habit of positive recognition forces you to be on the lookout for things to praise, which in turn causes you to pay attention to what individuals are bringing to various projects. You don't have to do this in public, but every week there should be at least one thing you can recognize about someone on your team. Even better, look for something to recognize weekly for everyone who reports to you.

3. **Provide lightweight, regular feedback.** Start with positive feedback. It's both easier and more fun to give positive feedback than it is to give corrective feedback. As a new manager, you don't have to jump into the deep

end of coaching first thing. Many people respond better to praise than they do to corrective feedback, and you can use kudos to guide them to better behavior by emphasizing the things they've done well.

Positive feedback also makes your reports more likely to listen to you when you need to give them critical feedback. When they believe that their manager sees the good things they do, they'll be more open to hearing about the areas where they might improve. It's best to give critical feedback quickly in the case of an obvious misstep, but continuous feedback is more than in-the-moment corrections. Use a habit of continuous feedback to talk about things that don't seem to be going well as you start to notice them, rather than waiting until the review cycle to have uncomfortable conversations.

Bonus: Provide coaching. Ultimately, continuous feedback works best when you, as a manager, pair that feedback with coaching. As situations arise, use coaching to ask people what they might have done differently. When things are going well, praise them, but also make suggestions as to what could be even better in the future. Coaching-based continuous feedback means going beyond a simple "good job" to really engage with the details and form a partnership with your direct report where the two of you are working together to help her grow.

Why do I list coaching as a bonus? It's not always a core need for doing the job well, and there will be many times when you don't have either the qualifications or the capability to provide the coaching that everyone on your team needs. Coaching is most important for your early-career team members, or those who have the potential or desire for advancement. Many people will be content with doing the job they know how to do well, and as long as they are doing it well enough, it's not a good use of your time to try to coach them. Save your valuable coaching time for those who are receptive to it.

Performance Reviews

Continuous feedback, even if it's just regular recognition for good work, is an important tool in the hands-on manager's toolkit. However, it's not a replacement for a more formal, 360-based performance review process.

The *360 model* is a performance review that includes feedback from, in addition to a person's manager, his teammates, anyone who reports to him, and coworkers he regularly interacts with, as well as a self-review. For example, an engineer with no direct reports might solicit reviews from two other engineers on her team, the new hire she was mentoring, and the product manager she works with. Performance reviews take a long time because you need to give and receive feedback from many different people. As a manager, you then have to gather all that feedback together and summarize it for the person being reviewed.

Performance reviews reward the time spent by providing a valuable chance to synthesize a bunch of information about a person. Beyond that, 360 reviews give you at least a high-level view into what other people are thinking about your direct reports. The self-reviews give you a sense of what your people believe about themselves, their strengths and weaknesses and accomplishments over the year. Writing the summary review gives you the chance to focus for longer than a few minutes on the individuals and look at the big picture over a longer period of time. All of this should help you see some patterns and trends that you might overlook in the process of day-to-day continuous feedback.

Performance reviews go wrong because people aren't given time to prioritize working on them, and many people find them hard to write. They go wrong because we tend to remember and overemphasize things that happened most recently, and forget about the things that happened six months or a year ago. They go wrong because we all suffer from various biases of which we may or may not be aware, and we tend to review people through the lens of those biases, criticizing some people for behaviors that we don't even notice in others. All of these things are true, and you will probably see all of them play out. Despite all that, this process is incredibly valuable, and you as a manager have the opportunity to make it more or less valuable depending on how you approach it.

WRITING AND DELIVERING A PERFORMANCE REVIEW

Here are a few guidelines for writing and delivering a successful performance review.

Give yourself enough time, and start early

This process isn't something you can knock out in an hour and do well. You have a million things on your plate, but plan to spend solid, uninterrupted time working on reviews. Work from home if you need to. You owe your team enough time to read the collected feedback, digest it, and summarize it well. My advice is to start by reading the collected reviews and taking a few notes, processing the information for a little bit before trying to write a full summary. Give yourself enough time to write and come back to what you've written at least once before you have to submit the review.

Most companies expect that managers will read feedback and anonymize it as part of writing up their summary, but some companies have open processes where the original peer feedback is visible and identifiable to the person being reviewed. Even in an open process, as the manager you should still read that feedback and use it as part of your review writing, since the manager review is still often considered the most important summary of all review feedback.

Try to account for the whole year, not just the past couple of months

This will be easier if you keep notes on what has happened with each person throughout the year. One tactic is to keep a running summary of your 1-1s, including any feedback that was delivered. If you haven't done this, I encourage you to look through your email to remember which projects launched, review what activities were happening month by month, and put yourself back into the perspective of that time period. The goal for viewing the whole year is to recognize not just early accomplishments but also the growth and change you've seen since then.

Use concrete examples, and excerpts from peer reviews

Anonymize peer reviews, if needed. If you can't use a concrete example to support a point, ask yourself if the point is something you should be communicating in the review. Forcing yourself to be specific will steer you away from writing reviews based on underlying bias.

Spend plenty of time on accomplishments and strengths

You want to celebrate achievements, talk about what's going well, and give plenty of praise for good work. This goes not only for the writing process but also—and especially—for the delivery. Don't let people skip over the good stuff in order to obsess over the areas for improvement, as many will want to do. Those strengths are what you'll use to determine when people should be promoted, and it is important to write them down and reflect on them.

When it comes to areas for improvement, keep it focused

Writing about areas for improvement is often a tricky part of the feedback. In the best case, there are a couple of clear themes that run through peer feedback, and that you have observed, to comment on. Here are some examples of themes that I have seen. There are people who:

- Struggle with saying no to distractions and end up helping with other projects instead of finishing their own
- Do good work but are hard for others to work with, tending to be overly critical or rude in meetings, code reviews, or other collaborative activities
- Struggle to break their work up into intermediate deliverables, and don't balance planning and design with getting things done
- Work well with other engineers but do not work well with other departments or teams
- Struggle to follow the accepted best practices of the team, cut corners, or otherwise do sloppy work

More often, you'll get a lot of scattershot feedback that's moderately helpful at best. Some people will seem to be reaching for something to say, and others will have a particularly harsh impression that no one else seems to share. Especially in the case of scattershot feedback, make sure that the feedback you're seeing makes sense before you deliver it. For example, if only one reviewer mentions sloppy work, is the problem that the work is sloppy, or that the reviewer has higher standards than the rest of the team? Use your judgment in this case. If the feedback seems valuable for the person to hear, share it, but don't just blindly report all grudges.

What about the case where you have very little meaningful feedback for improvement? This indicates that the person is ready to be promoted or given more challenging work. If the person is doing a solid job at her level but isn't ready for promotion, the feedback should indicate one or two skills she needs to expand to become qualified for promotion. Some people may never need to be promoted out of their current level, but the nature of the tech industry is such that skills need to be refreshed to stay current, so you can also focus on new technical learning opportunities.

Avoid big surprises

Set expectations appropriately before reviews are delivered. If someone is underperforming across the board, the review should not be his first time getting that feedback. Similarly, if someone has recently been promoted, you may want to prepare her for the fact that she will be reviewed based on higher standards.

Schedule enough time to discuss the review

I usually give people a printed copy of the review as they're leaving on the evening before the review is scheduled. This practice gives them a chance to read it at home, and then come to the meeting ready to talk about what it says. Even though they've had the review and gotten to read it, I still take the time to go over each section, starting with the strengths and accomplishments. Again, don't let them skip over this and jump straight into the areas for improvement. Many people are uncomfortable being praised at length, but skipping that section undermines its value in reinforcing and encouraging their talents.

Some reviews are summarized by a scaled ranking, such as a number from 1 to 5 or the equivalent in words ("fails to meet," "meets," "exceeds"). If you have to do this, expect it to be the hardest part of the review to discuss for anyone who got anything less than the top ratings. In my experience, people are uncomfortable being told they merely meet expectations, especially those who are early in their careers. Come prepared to dig into the reasons for this score, including examples of how the person could achieve a higher score.

Ask the CTO: Identifying Potential

Is there a good way to identify potential? Does all potential look the same? What does it really mean for someone to have potential?

People often make a critical mistake when it comes to understanding potential. They see it as a set of inborn traits, or something that can be determined purely from credentials. "He went to a great school, so he has high potential!" "She is very articulate, so she has high potential!" Or, the most blunt, "He is handsome and tall and male, so he has high potential!" Biases lead us to assume potential, and moreover, to give people the benefit of the doubt long past the point at which they've shown that their "potential" is an illusion.

I'll make a suggestion to all of you. A person who has never shown reasonable performance, and who has been with a company long enough for you to observe performance, probably doesn't actually have potential, at least within that company. It doesn't matter how good his school was, how articulate she is, how tall he is...if the employee has been with a company for a while with little to show for it, all that potential you are imagining is simply that—a figment of your imagination (or your biases).

Real potential shows itself quickly. It shows itself as working hard to go the extra mile, offering insightful suggestions on problems, and helping the team in areas that were previously neglected. The person who has potential but isn't yet showing equivalent performance is showing up for the team in a way that others do not, even if her work is slow-going. It's rare to see someone with true potential in a company and poor performance, though you may see slightly below middle-of-the-road performance. Often the solution to this problem is to move the person to a place where his potential can be realized. A person who has a strong visual design sense but is struggling with the day-to-day of completing coding tickets may do better in a UI/UX role. A great firefighter who hates planning may be better suited to an operations-focused team.

Don't confuse "potential" as it might be described by a grade-school teacher with the type of potential you care about. You are not molding young minds; you're asking employees to do work and help you grow a company. Potential, therefore, must be tied to actions and value produced, even if it's not directly the value you expected to see produced. The sooner you can get over the disappointment that a high-potential person didn't work out, the sooner you can identify the true high-potential stars on your team and develop them fully.

Cultivating Careers

One of my most critical promotions happened during my time in finance. The finance world has a strange way of giving out titles. Drawing from the days when firms were built on a partnership model, there tend to be only a few "public" titles: Associate, Vice President, Managing Director, and Partner. The Vice President title is a critical leap. Achieving it is (or was) a sign that a person has proven herself worthy of building a long-term career at the firm. Therefore, the time it takes you to get a VP title is a strong signal for your future success, and getting the promotion is a complex process that's done only once a year and run by the senior managers.

My manager explained this to me twice. First, when I got my own VP promotion, he walked me through all of the materials we'd be gathering to support my case. Projects shipped, yes, but also signs of leadership, and work that pushed me beyond my immediate team. The second time I went through this was when I prepared the packet for someone who worked for me. We gathered all sorts of evidence, including the letter the candidate got commending him for being the floor fire warden. Both of these promotions were successful, but I have no doubt that we succeeded at least partially because my boss/mentor knew exactly how to play the game.

If you're a manager, you are going to play a key role in getting people on your team promoted. Sometimes it will simply be up to you to determine who gets promoted, but more commonly promotions will be reviewed by your management, or a committee. So you'll not only need to have a good idea about who deserves to be promoted, but you'll need to make a case for their promotion as well.

What does this process typically look like? Generally, you'll look at the people on your team a couple of times a year, consider their job level, and ask yourself, are any of these people close to the next level? In the case of the early-career staff, the answer is likely to be yes. These days, people fresh out of college tend to get promoted at least once in their first couple of years on the job, because they're often hired in at an "up or out" level.

To clarify, take the example of Famous BigCo. Famous BigCo hires engineers out of college at level E2 (level E1 is reserved for interns). Famous BigCo has a policy that an engineer who shows no sign of advancing past level E2 after two years at that level doesn't have a future at the company. It has this policy for levels E2–E4, but at E5, you can stay forever.

So, if you have a team of E2s and E3s, you need to be preparing them to be promotable every couple of years. Fortunately, this is usually straightforward. As long as you don't stop them from getting promoted, they'll be moved forward by the process. Your job with this group is to make sure that they're learning how to estimate their own work, getting it done roughly within the estimates, and learning from their mistakes. The evidence for promotion often takes the form of projects or features they've completed independently, participation in on-call rotations or other support, and engagement in team meetings and team planning.

The important thing for you to start doing now that you're in management is to learn how the game is played at your company. Every company has its own variation of the promotion process, and you're probably in this role because you survived it. If you don't know how it's done, ask your manager for advice. How are these decisions made? How early do you need to start preparing packets? Are there limits on the number of promotions that can happen in any given year? As you learn how to play the game, I encourage you to be fairly transparent with your team. When members express the desire to be promoted and they don't have a strong case for promotion, telling them what goes into the process will help them understand what they may need to change.

You should also prepare yourself to start identifying promotion-worthy projects and trying to give those projects to people who are close to promotion. You, as the manager, are in a good position to identify what's coming up for the team. Depending on how work gets assigned, you may either directly assign these projects to people, or encourage people to volunteer for projects that are a stretch goal for them. Keep an eye out for opportunities for your team members to stretch themselves and grow.

This work does start to change the more senior your team becomes. Many people will not continue to advance past a certain level, at least not within the same company or team. There are fewer opportunities for people to show the kind of leadership or breadth of impact needed to get promoted as they become more senior. Sometimes there is nothing you can do about this, except perhaps to refer them to other leaders in different parts of the company for mentoring or guidance. As much as it might hurt you to lose them, they may be better off in another team or even another company with new challenges.

Many companies expect you to be acting at the next level before you get promoted to it. This practice exists to prevent the "Peter Principle," in which people are promoted to their level of incompetence. It also signals that there's room for another person acting at that level on the team. Keep this in mind as you think about your team's careers. If there is no growth potential on your team because there's no room for people to work at a more senior level, it may be a sign that you need to rethink the way work is done in order to let individuals take on bigger responsibilities.

Challenging Situations: Firing Underperformers

One of the hardest things that any manager must do is to fire someone for underperformance.

This is difficult to write about because so much of the act of firing employees is dictated by HR departments these days, even at small companies. There are good and bad things that come out of this, but arguably the nicest thing is that you, as a manager, will have a process and procedure to follow. Upon hearing that someone is underperforming, many companies will have you write the person a document called a *performance improvement plan*. This is a set of clearly defined objectives that the person must achieve within a fixed period of time. If she manages to achieve them, then she is taken off the plan and all is well; otherwise, she's fired. Depending on the company, such a plan might actually be an effort to turn an employee around, but often the plan is written in such a way that the person can't possibly hope to achieve the goals in the allotted time, and it's just a generous way of giving someone time to look for another job before being fired.

Whatever the procedure is at your company, the process of coaching someone out should begin long before any performance improvement document is filed with HR, and long before the actual act of firing. One of the basic rules of management is the rule of no surprises, particularly negative ones. You need to understand what a person is supposed to be giving you, and if that isn't happening, make it clear to her early and often that she is not meeting expectations.

The ideal is that you know exactly what job she is supposed to be doing, and if she isn't doing it, you can say, "You aren't doing X, Y, and Z. Do more of those things." Of course, like all perfect circumstances, reality is rarely so simple.

A common, straightforward scenario is closer to the following. Your employee, Jane, has been with you for a few months. She seemed a little bit slow in the onboarding process, but you gave her the benefit of the doubt; the code base isn't in perfect shape, and there's a lot of business jargon to learn in a new hire's first few months. However, it's been six months, and when you look back over that time, you see very little in the way of achievement on Jane's part. In fact, the few things she has done have not gone well—they have been very late, very buggy, or both.

This situation sounds straightforward on paper. Tell Jane that she is not meeting expectations, that her work is too slow or not well done, and give her a strict deliverable. But Jane has excuses, of course, and some of them are believable. The onboarding wasn't good. Her first month was interrupted by the company party, and then you were out for a week on vacation, and she has not had anyone to ask questions of. In fact, it sounds a little bit like the issue is you and the team, not her at all.

This situation is why you start giving feedback early and often, and keep records of the feedback you've been delivering. Feedback, positive or negative, should be a conversation. If you avoid tackling negative feedback until it builds to a boiling point, you're going to be met by a pile of excuses, and then what do you do? Some managers will ignore the excuses at their peril, and lose employee after employee to an unwelcoming team that fails to onboard, coach, and give clear goals to employees. On the other hand, some managers will accept any excuse until problems can no longer be swept under the rug, and the team is furious at management's inaction with regard to the lagging employee.

You'll always need to have a record of negative feedback to fire someone in any environment where HR is active and a standard performance improvement plan is required. If you have no HR, I suggest that you still give people clear improvement feedback in writing, with a timeline for improvement, and have them acknowledge it in writing as well (email is OK). Not only does this protect you legally, but it helps you treat your employees fairly.

A final warning: don't put anyone on a plan whom you wouldn't be happy to lose. Most smart employees will take this formal warning as a sign that the organization is not a good fit for them, and leave as quickly as possible. I once heard a story about a great engineer who was put on a surprise performance improvement plan by his manager after someone in the organization complained that he had bowed out of a project. This manager, who had not been paying any attention to the situation and had approved the engineer focusing elsewhere, gave in to pressure to set up a plan that served to do nothing but poison any goodwill the engineer might have had for his manager and the company. It's no surprised that the engineer resigned soon thereafter, despite having easily achieved the goals of the improvement plan.

Ask the CTO: Coaching Someone Out of the Company

I have an employee who seems stuck. He's been with the company for a couple of years and done OK work, but I don't think he has the potential to be promoted further on our team. Every time he asks what he needs to do to get to the next level, I tell him, but he then goes back to his comfort zone, and no amount of nudging seems to bring about any change. What should I do?

This is a fairly common occurrence that managers have to handle. You have an employee who has topped out in the organization and seems to be losing energy. He has achieved expectations in his level, but can't figure out how to grow enough to get to the next level, despite your efforts. It may be time to coach him out.

Many organizations have a rule of "up or out" for early-career employees. The entry levels of most engineering career ladders expect that people at those levels will progress within a certain period of time, and if they don't, they aren't meeting expectations and will be fired.

Generally, you want to make sure that long-term employees are capable of doing their day-to-day work independently, without a lot of oversight or help. However, once people have gotten past these up-or-out career points, what do you do when they get stuck?

Some people will be happy cruising as senior engineers or managers at a certain level for their whole careers, and if you're both satisfied with the work, there's nothing wrong with that. Others, like your employee, want to progress but for whatever reason don't seem to be able to do it on your team. You owe it to your employee to be clear that this is the case. This is what is meant by "coaching out." Make the situation clear to him. You have told him repeatedly what the next level looks like, and he has not been able to show that he can work at that level, so you don't think that your team is the right place for him to grow his career. You aren't firing him, but you are telling him that he needs to move on if he wants to progress.

Give the employee a chance to find a job in another part of the organization or at another company. When he does, let him go happily, and do your best to retain goodwill. Former couples who break up because they don't see a future together can remain friends, and the same goes for former employees who simply need a different team or company to shine.

Assessing Your Own Experience

- Have you set up regular 1-1s with your direct reports?

- When was the last time you talked to your reports about their career development? If it was more than three months ago, can you make sure to put this in your next 1-1s?

- Have you given feedback to your reports in the last week? When was the last time you handed out kudos in front of the team?

- When was the last time someone behaved in a way that needed correction? How long did it take you to give corrective feedback? Did you give the feedback in private, or did you do it in public?

- Have you ever been given a performance review that felt like a waste of time? What was it missing that could have made it more valuable?

- What was the most useful piece of performance feedback you ever got? How was it delivered to you?

- Do you know how the process of promoting people works in your company? If not, can you ask someone to walk you through it?

5

Managing a Team

It's a short step from managing a person or two to managing a whole team, but managing a team is more than just doing the job of managing the individuals. At this point, your job has changed. In fact, at every step beyond this level you will probably experience a totally different set of requirements and challenges. The hardest thing to prepare for as you advance in your career is the idea that you're going to start doing totally different things. As much as you may want to believe that management is a natural progression of the skills you develop as a senior engineer, it's really a whole new set of skills and challenges.

Here is the job description that I used for the role of managing a team, which I called "engineering lead":

> *The engineering lead will spend less time writing code, but they still engage in small technical deliverables, such as bug fixes and small features, without blocking or slowing down the progress of their team. More than writing code, they hold responsibility for identifying bottlenecks in the process and roadblocks to success for their team and clearing these roadblocks.*
>
> *The person who fills this role is expected to have a large impact on the success of [the organization] as a whole. In particular, leaders in this role are capable of identifying the most high-value projects and keeping their team focused on these projects. As part of keeping the team focused, the engineering lead will partner closely with the product lead to manage project scope and ensure the technical deliverables are met. In addition to focusing the team, they are capable of identifying headcount needs for the team and planning and recruiting to fill these needs.*

The engineering lead is an independent manager. They are comfortable managing team members with different skill sets from their own. They communicate expectations clearly to all team members, and solicit and deliver individual feedback frequently (not just in the scope of review periods). In addition to strong management skills, the engineering lead acts as a leader for the technical roadmap for their product group (pillar). They clearly communicate the timeline, scope, and risks to their pillar partners, and lead the delivery of major initiatives on clear timelines. Additionally, they identify areas of strategic technical debt, do the cost/benefit analysis for resolving this debt, and communicate suggested timelines for prioritizing this to the management team.

We've covered the basics of managing individuals, but now let's talk about what it takes to lead a whole team, all while still being technical.

The theme of this chapter is a focus on the job beyond the people management. Because it's easy for new managers to get overly focused on the people-related tasks, I want to draw your attention back to the more technical, strategic, and leadership areas of managing a team.

Becoming a People Manager

I started out as an informal team lead in a company that had resisted traditional managers. After I played that role for a while, it came time for me to become an official people manager. Managers as a role were new to the company, and the entire organization met the changes with some trepidation. When dividing engineering among managers, we weighed who would chafe at the new structure. Not everyone was comfortable working for their former peers, but I was pretty lucky. Most of the people I now managed had worked with me long enough to be OK with the idea that I was now their manager. Their support helped tons. It wasn't perfect, but the pushback I received wasn't significant.

In this new role, I found myself managing a few people who were far more senior, tech-wise, than I was. It was the first time that I couldn't rely on having the most knowledge as my main leadership tool. This wasn't simple impostor syndrome. I knew I was out of my league. *They* knew I was out of my league! Of course, both of the most senior engineers I was now managing realized that this was awkward. We talked about how everyone had a job to do, and mine was to help them succeed however I could.

One engineer continued to help me along the way with ongoing feedback. I worked hard to learn what was important to this person and what they needed to succeed. The other engineer had trouble adjusting to me being his manager. He transferred to another team—at first. Months later, a bit contrite, he returned to our team and agreed to work with me. It turns out, being a good manager isn't about having the most technical knowledge. The work of supporting people was far more important to management success.

—*bethanye Blount*

Staying Technical

This book is for engineering managers. It's not a generic management book. Engineering management is a technical discipline, not just a set of people skills. As you progress in your career, even though you may stop writing code, your job will require that you guide technical decision making. Even with architects who design the systems or other senior technical staff who are in charge of the details, as the manager of a team, you have the job of holding those people accountable for their decisions, of making sure that the decisions pass the technical smell test and have been balanced against the overall context of the team and the business. Technical instincts honed over years of doing the job are very important for guiding that process.

Furthermore, if you truly wish to command the respect of an engineering team, they must see you as technically credible. Without technical credibility you face an uphill battle, and even though you may be able to get into a position of leadership in one company, your options will be limited. Don't underestimate the value of your technical skills as you work to become a successful engineering manager.

Of course, you have to learn how to balance. It's a struggle to figure out how to stay technical as you transition to management. The new responsibilities that come with being a manager—more meetings, planning, administrative tasks—don't lend themselves to having focused time to write code. It can be hard to work out how to stay in the code when you are pulled in a million directions.

However, at this level, if you don't stay in the code, you risk making yourself technically obsolete too early in your career. You may be on a management career path, but that doesn't mean that you should wash your hands of technical responsibilities. In fact, I mention specifically in my engineering lead job description that I expect managers at this level to implement small features and bug fixes.

Why bother writing any code if all you're doing is small stuff? The answer is that you need to stay enough in the code to see where the bottlenecks and process problems are. You might be able to see this by observing metrics, but it's far easier to feel these problems when you're actively engaged in writing code yourself. If the build is really slow or deploying code takes too long or on-call is a nightmare, you'll feel it in the difficulties you, an experienced engineer, have in knocking out trivial programming tasks. Imagine how frustrating it is for the members of your team! It's far easier to identify technical debt and prioritize dealing with it when you've slogged through the code yourself.

Additionally, as the manager of a single team, you'll be called upon to help guide what is possible and impossible to do in your systems. When the product manager for your group has a crazy idea, it's much easier to manage when you're confident in your ability to evaluate how easy that feature will be to implement in the given systems (although beware of overconfidence in giving these estimates!). Strong engineering managers can identify the shortest path through the systems to implement new features. As you learned in your time as a tech lead, a critical part of complex project management is understanding the pieces of the system well enough to determine the best path to implementation. The more you understand the code in the system, the easier determining this path will be.

Sadly, some companies don't really have the role of "manager who has a little time to code" available. These companies split the management and technical tracks so cleanly that managers immediately start with large teams reporting directly to them. Thus the manager's job becomes an administrative and people management position, and these managers end up grabbing technical time on nights and weekends, if ever. If your company is like this, my advice is to *stay technical* until you feel that you have truly mastered what you want to learn for writing code and designing systems, and then decide if you want to switch careers into management. It's hard to make up lost time when you stop writing code, and if you do it too early in your career, you may never achieve sufficient technical savvy to get beyond the role of middle management.

If you are horrified at my suggestion that managers stay in the code at all, don't worry! In later chapters I'll talk in detail about the point at which it doesn't make sense to be in the code anymore, and I do believe that point exists. But for now, try to stay in it a little bit. I promise, it'll make your job easier.

Debugging Dysfunctional Teams: The Basics

Sometimes you'll find yourself managing a dysfunctional team. They keep missing deliverables. People are unhappy. They keep quitting. The product manager is frustrated. The team is frustrated with the product manager. Or maybe they're just lacking energy in their work, or lacking enthusiasm about the current projects. You can tell something is wrong, but you're not entirely sure what it is. A few basic dysfunctions can creep into tech teams. I'll introduce these dysfunctions briefly here so you know what to look for and how to solve it.

NOT SHIPPING

You may not think this is a dysfunction. Perhaps your team is in deep research mode on a new problem, for example. However, even teams doing research generally have goals and deliverables, even if they're just in the form of initial findings. Humans, by and large, feel good when they set small goals and meet them regularly.

As the manager of the team, you may worry about pushing them too hard, and so you let them miss deadlines without a fuss. The trick is to learn how to balance pushing your team and holding back. If you're still writing code for the team, this may be a good time to roll up your sleeves and help the team meet its deliverables, or really dig in to the part of the project that's slipping and partner with the engineers responsible to help understand the situation.

Sometimes, teams aren't shipping because the tools and processes they've been using make it hard to get work done quickly. A common example is that your team only tries to release changes to production once a week or less. Infrequent releases can hide pain points such as poor tooling around releases, heavily manual testing, features that are too big, or developers who don't know how to break their work down. Now that you're managing the team, start to push for the removal of these bottlenecks.

At my last job, there was a critical part of the system that, for a time, we released only once a week. The releases took hours and were very painful, and often suffered from people trying to get in last-minute changes that broke tests and slowed everyone down. We all decided this was a problem and the team came together to improve the code base and automation, in order to make releases happen faster. Toward the end of the process, I pushed the team to make improvements that allowed us to release daily. The impact of this change on the team was immediate. It turns out that releases can be a point of resource contention. When people are contending for a scarce resource, conflicts and unhappiness among team members are almost inevitable. Making the code-shipping resource far less scarce immediately improved team morale.

PEOPLE DRAMA

Sometimes we let ourselves hang onto that brilliant asshole for too long. You know, that person you think can't be replaced because he's just so productive and so smart, but who isn't a team player and makes everyone around him unhappy. (For more on this kind of toxic employee, see "The Brilliant Jerk" on page 90.) A less critical version of this situation is the person who just stirs up drama, who dwells on negative experiences, or who spends a bit too much time on gossip and playing games of us-against-them.

You have to be brave and nip people drama in the bud quickly. It's OK to ask your manager for help with this, especially if it's your first time doing it, but be aware that your manager may actually have an even harder time dealing with the brilliant jerk than you do. She isn't seeing the immediate impact on team dynamics; she's just seeing someone who gets things done. Be prepared to have a series of conversations with both the employee and your boss. It may be that a move to a different team will clear up the situation.

The negative person is easier to deal with than the brilliant jerk. Make it clear to him that the behavior has to change, bring clear examples, and provide corrective feedback quickly after things happen. Sometimes the negative person is just unhappy and the best thing to do is to help him leave the team on good terms; you must be prepared for this outcome. Other times, the person has no idea about the impact he's having on the team, and a quick chat will be all that's needed to curtail the incidents.

Be careful that vocally negative people don't stay in that mindset on your team for long. The kind of toxic drama that is created by these energy vampires is hard for even the best manager to combat. The best defense is a good offense in this case, and quick action is essential.

UNHAPPINESS DUE TO OVERWORK

This problem is much easier to solve. Usually, unhappiness due to overwork has a root in problems that you can address. For example, if overwork is due to (in)stability of the production systems, it's your job as the manager to slow down the product roadmap in order to focus on stability for a while. Make clear measures of alerts, downtime, and incidents, and strive to reduce them. My advice is to dedicate 20% of your time in every planning session to system sustainability work ("sustainability" instead of the more common "technical debt").

In a case where overwork is due to a pressing, time-critical release, remember two things. First, you should be playing cheerleader. Support the team however they need supporting, especially by helping out with the work yourself. Order dinner. Tell them you appreciate the hard work. Make it clear that they'll have explicit break time after the push. Make it as fun as you can in the moment. Sometimes a crunch period can serve as a bonding experience for a team. But they'll remember whether their manager was with them during the stressful period, or off somewhere else, doing her own thing.

Second, do everything you can to learn from this crunch period and avoid it the next time. Cut features if you can. Push back on the date if it's truly unrealistic. Crunch periods will happen, but there is no reason they should happen frequently.

COLLABORATION PROBLEMS

Your team isn't working well with the product team, or the design team, or another tech team, and the lack of collaboration is dragging everyone down. There's no quick fix here, but showing a willingness to improve collaboration goes a long way. If you aren't already, make sure you're having regular touchbases with the appropriate peers to work through issues. Gather actionable feedback from your team, and have productive conversations about possible improvements. You can make the situation worse by undermining your peers in front of your team, so even when you are frustrated with them, try to stay positive and supportive of their efforts in public.

If your team isn't working well together, look into creating some opportunities for them to hang out without it being all about work. Taking the whole team to lunch, leaving work early on a Friday afternoon to attend a fun event together, encouraging some PG-rated humor in chat rooms, and asking people how their lives are going are all ways to cultivate team unity. As a new manager I was pretty reluctant to get into this type of bonding, but even most introverts want to have a feeling of relatedness with their team. Assuming you don't have any of the "people drama" problems listed earlier, small efforts in this area can warm the group up considerably.

Ask the CTO: Managing a Former Peer

I just got promoted to run my team over another peer of mine, a senior engineer who also wanted the job. How can I make sure to manage this so that I don't alienate him while still taking on the role?

This experience can be deeply awkward, so the first thing to do is to acknowledge that. If you're now managing someone who was truly your peer, acknowledge the weirdness of the transition. Be honest and transparent with this person that you're going to do the best job you can, but you'll need his help to do it. You need him to be honest with you about the things that are going well and the things that are not. You're going to have to be a little bit vulnerable with him, because you won't be perfect the first time around.

Next, remember that your job has changed in some big ways. As his manager, you may now have the ability to override his decisions, but use this power very cautiously. Using your managerial power to override technical decisions is usually a bad idea. Resist the temptation to micromanage people—especially those who used to be your peers. They're going to be sensitive to the feeling that you've been "rewarded," even if they didn't want to become managers themselves. If you question their every move and try to make every single decision yourself, you'll make this sensitivity much worse.

A corollary here is that you're going to have to let go of some of your previous work as you ease into the additional responsibilities of people management. Every step up the management chain will mean adding new responsibilities and giving up some of your old ones. You can use this situation to your advantage with former peers by openly giving them more control over some of that technical work you used to own. This is also an opportunity to give new challenges to more junior members of the team. While many engineering organizations want first-level managers to continue to write some code, they likely expect the code those managers write to be smaller features, bug fixes, and enhancements, rather than deep new systems.

Throughout all of this change, your goal is to show the team that you're committed to helping them succeed. Your new role isn't taking anything away from the rest of the team; it's only giving you some new responsibilities that either were being neglected or used to belong to someone else, and shifting some of your old responsibilities to other members of the team.

Your team won't be successful if your former peers all quit because they can't stand working for you. They're going to be extra-sensitive to any disagreements or perceived power grabs on your part. They may even do things to try to undermine you. Pick your battles. In the long run, handling this transition with maturity will pay off.

The Shield

Many pieces of management advice tell new managers that part of their job, if they are effective, is to be a shield (or, less politely, a "bullshit umbrella"). They should help their team focus on what they need to get done without being distracted by the wider drama, politics, and changes happening in the company around them.

I have mixed feelings about this take on management. I do think that teams who are unnecessarily exposed to toxic drama that doesn't concern them tend to get distracted and stressed out. If you're managing an engineering team, they don't need to be concerned with interpersonal incidents in the customer service organization. I've watched with mixed pride as my own teams continue to function smoothly when it seems to me like the world is burning down around my ears. It's valuable for everyone to realize that they can and should focus on the things they can impact and change, and ignore the things they can't. Drama in the workplace is usually little more than an ego-entertaining drain.

So, yes, shielding your teams from distraction is important. Or, to put it another way, helping them understand the key important goals and focusing them on those goals is important. However, it's unrealistic to expect that you can or should shield your team from everything. Sometimes it's appropriate to let some of the stress through to the team. The goal is not to stress them out, but to help them get context into what they're dealing with. The extreme shielders think they can best focus and motivate their teams by giving clear goals. But humans usually need some sort of context into *why* these goals have been set, and thereby into what problems they're working to solve. If you're going to have operational issues in November if a particular system isn't up and running, your team deserves to understand that consequence. Appropriate context is what helps teams make good decisions about how and where to focus their energy. As the manager, it's not your job to make all of those decisions by yourself.

Another error that the shield sometimes makes is denying that any drama exists in the outside world. If layoffs happen in another part of the company and the team finds out from someone else, rather than shielding your team from drama, you've created a situation where they feel like something bad is happening and no one wants to admit it. If you instead communicate information about such events in a straightforward, low-emotion way, you alleviate the gossip and quickly neutralize the impact on your team.

You may be a shield, but you are not a parent. Sometimes, in combining the roles of shield and mentor we end up in a parenting-style relationship with our team, and treat them like fragile children to be protected, nurtured, and chided as appropriate. *You are not their parent.* Your team is made up of adults who need to be treated with appropriate respect. This respect is important for your sanity as well as for theirs. It's too easy to take their mistakes personally when you view them as a child-like extension of yourself, or to get so emotionally invested that you take every disagreement they have with you personally.

How to Drive Good Decisions

What is your role in the decision-making process for your team? Do you know? You may have a product manager who works with your team and owns the product roadmap, or the set of business features that your team has committed to working on. You probably have a tech lead who, as we covered in Chapter 3, is still deep in the technology but is also thinking about project management and the work that needs to be done. So where does that leave you, the engineering manager?

You have more responsibility than you may expect. While the product manager is responsible for the product roadmap, and the tech lead is responsible for the technical details, you are usually accountable for the team's progress through each of these elements. The nature of leadership is that, while you may only have the authority to guide decisions rather than dictate them, you'll still be judged by how well those decisions turn out.

CREATE A DATA-DRIVEN TEAM CULTURE

When you have a product or business head, she should be accustomed to using data about the business, the customers, the current behavior, or the market potential to justify her decisions. Start adding other data to the mix. For example, give that person data about team productivity (such as the time it takes to complete features) or data about quality measures (like how much time is spent dealing with outages, or the number of bugs found in QA or after releases). These efficiency and technical data points can be used to evaluate decisions on both product features and technical changes.

FLEX YOUR OWN PRODUCT MUSCLES

Strong leadership cares about cultivating success and having a team that delivers successful projects, which means honing your understanding of what is important to your customer. Whether you're writing code for an external customer, developing tools for other engineers, or even running a support team, you have some group that depends on the output of your work. Treat them as your customers. Taking the time to develop customer empathy is important because you'll need to give your engineers context for their work. Developing customer empathy will also help you figure out which areas of the technology have the greatest direct impact on your customers, and that understanding will guide where you invest engineering effort.

LOOK INTO THE FUTURE

You need to think two steps ahead, from a product and technology perspective. Getting a sense of where the product roadmap is going helps you guide the technical roadmap. Many technical projects are supported on the strength of their ability to enable new features more easily—for example, rewriting the checkout system to plug in payment types like Apple Pay, or moving to a new JavaScript framework model that supports streaming data changes via WebSockets, in order to build a more interactive experience. Start asking the product team questions about what the future might look like, and spend some time keeping up with technological developments that might change the way you think about the software you're writing or the way you're operating it.

REVIEW THE OUTCOME OF YOUR DECISIONS AND PROJECTS

Talk about whether the hypotheses you used to motivate projects actually turned out to be true. Was it true that the team moved faster after you rewrote that system? Did customer behavior change in the way the product team predicted when you added the new feature? What have you learned from your A/B tests? It's easy to forget to review assumptions after the project is done, but if you make this a habit for yourself and your team, you'll always learn from your decisions.

RUN RETROSPECTIVES FOR THE PROCESSES AND DAY-TO-DAY

Agile processes usually have a retrospective meeting at the end of each two-week development sprint, where you discuss what happened during the sprint and pick a few events—good, bad, or neutral—to discuss in detail. Whether you work in an agile methodology or some other fashion, the regular process retrospective has a lot of value for detecting patterns and forcing a reckoning with the outcome of decisions. Is the team feeling good about how they get requirements? Do they feel good about the code quality? This process helps you learn how the decisions you make over time affect the way your team operates in the day-to-day. This approach is more subjective than gathering data about the team's health, but it's arguably even more valuable than many objective measures, because it comes from the things the team itself is noticing and struggling with or celebrating.

Good Manager, Bad Manager: Conflict Avoider, Conflict Tamer

Jason's team is overworked. Everyone knows that Charles should be working on the big system rewrite, but he's been off on his own pet project for months. After hearing complaints that Charles isn't helping with the new system, Jason calls the team together and asks them to vote on what projects should be dropped to get the workload down. It's no surprise to anyone that they vote to drop Charles's pet project—no surprise, that is, to anyone except Charles, who has never heard anything about this from Jason and who figured he was doing the right thing.

Jason's team is feeling the crunch partly because Jason doesn't seem to stand up for them with other teams. He hates to say no to new projects, but he also doesn't ask for more people to help manage the load. Jason is cool, everyone agrees, but it is so hard to get him to actually act on resolving conflicts or making difficult decisions. As a result, the team is overworked, struggles to prioritize a way forward, and is nursing several grudges among its members.

Lydia's team is also feeling crunched, and she has her own Charles to deal with. She promised Charles that he would get time to work on this project, but it's clear that priorities have changed and so his work will need to change with them. In her 1-1 with Charles, Lydia explains the current workload, and tells Charles that his team needs him to help with the system rewrite. Charles is unhappy, and Lydia doesn't enjoy this conversation, but she knows that as the manager of the team, she's responsible for making sure they're focused on the most important projects.

Lydia knows that this project is important for the team to own, so while she pushes for more people, she makes sure that the team knows why she decided to take on this big project. She works with the team to prioritize the work, and guides them through their disagreements on what technology to use by following a structure for presenting options and soliciting feedback. Lydia's team describes her as tough but fair, and though disagreements happen, the team is good at getting through challenges and collaborates well.

When put in such stark terms, it seems pretty clear that Jason is not handling conflict well, while Lydia is taming it. While it seems like Jason's democratic style should lead to an empowered team, his inability to say no or to take the responsibility for any decisions means that no one feels very secure. It's hard to know what's going to happen next on Jason's team because instead of guiding the team, he's having the team guide itself.

Having a team that is constantly bickering and disagreeing is painful, and can be very dysfunctional. But there is such a thing as artificial harmony, and conflict-avoidant managers tend to favor harmony above functional working relationships. Creating a safe environment for disagreement to work itself out is far better than pretending that all disagreement does not exist.

THE DOS AND DON'TS OF MANAGING CONFLICT

- **Don't rely exclusively on consensus or voting.** Consensus can appear morally authoritative, but that assumes that everyone involved in the voting process is impartial, has an equal stake in the various outcomes, and has equal knowledge of the context. These conditions are rarely met on teams where each person has different levels of expertise and different roles. As when the team voted down Charles's work, consensus can be downright cruel. Don't set people up for votes that you know will fail instead of taking the responsibility as a manager of delivering that bad news yourself.

- **Do set up clear processes to depersonalize decisions.** When you want to allow for group decision making, the group needs to have a clear set of standards that they use to evaluate decisions. Start with a shared understanding of the goals, risks, and the questions to answer before making a decision. When you assign the ownership for making a decision to someone on the team, make it clear which members of the team should be consulted for feedback and who needs to be informed of the decision or plan.

- **Don't turn a blind eye to simmering issues.** Another way that conflict avoidance manifests is an inability to address problems until they've gone on for way too long. As a manager, if you're giving negative feedback in the course of a performance review, it shouldn't be a major surprise to your employee. There may be nuances that you didn't think through until writing the review, but if there are major problems with someone's work, that person should know about them as soon as you notice them. If you don't notice these problems yourself but learn about them during the review process via feedback from several peers, that's not a good sign. It's probably an indication that you are not paying attention, and not making space in your 1-1s for your team to discuss problems they're having with their colleagues.

- **Do address issues without courting drama.** There's a difference between addressing conflict and cultivating dysfunction. You want to allow space for people to express frustration, but mind the difference between letting off steam and a real interpersonal issue. Use your judgment as to what should be addressed and what should be dropped. The key questions to ask are: Is this an ongoing problem? Is it something you've personally noticed? Is this something many people on the team are struggling with? Is there a power dynamic or potential bias at play? The goal is to identify problems that are causing the team to work less effectively together and resolve them, not to become the team's therapist.

- **Don't take it out on other teams.** Ironically, conflict-avoidant managers often seek conflict when it comes to other teams. They identify strongly with their own team and will aggressively react to what they perceive as threats from outsiders. When something goes wrong, like an incident that spans across teams, the manager turns into a bully and demands justice for his team, or blames the problems on the other team. Sometimes, this behavior is an outlet for the manager's suppressed feelings about his own team. As one friend put it, "I was not telling my people the 10% of things they needed to be improving because I was afraid they would miss the message about the 90% of things that were good, and so I took that desire for accountability out on other teams. I really just wanted everyone to be fully accountable, and I needed to figure out how to express that internally and externally in a healthy manner."

- **Do remember to be kind.** It's natural and perfectly human to want to be liked by other people. Many of us believe that the way to be liked is to be seen as nice—that niceness is itself the goal. Your goal as a manager, however, should not be to be *nice*, it should be to be *kind*. "Nice" is the language of polite society, where you're trying to get along with strangers or acquaintances. Nice is saying "please" and "thank you" and holding doors for people struggling with bags or strollers. Nice is saying "I'm fine" when asked how you are, instead of "I'm in a really crappy mood and I wish you would leave me alone." Nice is a good thing in casual conversation. But as a manager, you will have relationships that go deeper, and it's more important to be kind. It's kind to tell someone who isn't ready for promotion that she isn't ready, and back that up with the work she needs to do to get there. It's unkind to lead that person on, saying "Maybe you could get promoted," and then watch her fail. It's kind to tell someone that his behavior in meetings is disrupting the group. It's awkward, and uncomfortable, but it's also part of your job as his manager to have these difficult conversations.

- **Don't be afraid.** Conflict avoidance often arises from fear. We're scared of the responsibility of making the decision. We're afraid of seeming too demanding. We're afraid people will quit if we give them uncomfortable feedback. We're afraid people won't like us, or that we'll fail when we take this risk. Some fear is natural, and being sensitive to the outcomes of conflict is a wise habit.

- **Do get curious.** Thinking about your actions is the best way to combat fear of conflict. Am I pushing this decision to the team because they really are the best people to decide, or am I just afraid that if I make an unpopular but necessary decision people will be mad at me? Am I avoiding working through this issue with my peer because she's truly difficult to work with, or am I just hoping that the issue will resolve itself because I don't want to have to discuss it and possibly be wrong? Am I holding back on giving my employee this feedback because he really was having a bad day and it's just a one-off, or am I holding back because I'm afraid he won't like me as a manager if I tell him? Be thoughtful about your behavior, and it's unlikely that you'll seek out unnecessary conflict.

Challenging Situations: Team Cohesion Destroyers

One of the critical elements of creating functional teams is building teams that work well and happily together. I was once given a test of a happy engineering team: "If you buy them pizza in the evening, will they stick around and socialize together, or will they race out the door as quickly as possible?"

I have some quibbles with that. Employees with obligations that take them out of the office at a strict time every day are no more or less engaged than those who are willing to stand around and chat. The larger point, however, is still a good one. Most gelled teams have a sense of camaraderie that makes them joke together, get coffee, share lunch, and feel friendly toward one another. They may have obligations they respect, and passions outside of work, but they don't view their team as something they're eager to escape every day.

The real goal here is psychological safety—that is, a team whose members are willing to take risks and make mistakes in front of one another. This is the underpinning of a successful team. The work of gelling a team begins by creating the friendliness that leads to psychological safety. You can encourage this by taking the time to get to know people as human beings and asking them about their extracurricular lives and interests. Let them share what they feel comfortable sharing. Ask how their child's birthday party went, how their ski trip was, how their marathon training is going. This is more than empty small talk; it fosters relatedness, the sense of people as individuals and not just anonymous cogs.

Beyond you personally cultivating relatedness, you want your teams to have their own relatedness among themselves. When companies talk about hiring for "culture fit," they often mean they want to hire people they can be friendly with. While this can have some unwanted consequences, such as discrimination, it comes from a wise place. Teams that are friendly are happier, gel faster, and tend to produce better results. I mean, do you really want to go to work every day with a bunch of people you hate?

This is why those who undermine team cohesion are so problematic. They almost always behave in a way that makes it hard for the rest of the team to feel safe around them. We refer to these employees as "toxic" because they tend to make everyone who comes into contact with them less effective. Dealing with them quickly is an important part of managing well.

THE BRILLIANT JERK

One variant of the toxic employee is the brilliant jerk, who, as we discussed earlier, produces individually outsized results, but is so ego-driven that she creates a mixture of fear and dislike in almost everyone around her. The challenge of the brilliant jerk is that she's probably been rewarded for a very long time for her brilliance, and she clings to it like a life raft. Acknowledging that there is value in the world beyond sheer intelligence or productivity would challenge her place in the world and tends to be a scary proposition for her. So she bullies with her intellect, cutting down dissenting voices in a harsh way, ignoring those she believes are not her equal, and letting her frustration with anything she sees as stupid seep out openly.

These days, most places claim that they don't tolerate brilliant jerks, but I personally don't believe that is true. It's incredibly hard for a manager to justify getting rid of someone who produces great work, even though she's a drain on everyone around her—especially if this person is only irregularly a jerk. How much jerk is too much? You start to run rings around the idea to try to justify keeping her on. You give her feedback, and she gets a little bit better for a while, but then she gets worse.

The best way to avoid brilliant jerk syndrome is to simply not hire one. Once they're hired, getting rid of brilliant jerks takes a level of management confidence that I think is uncommon. Fortunately, these folks will often get rid of themselves, because even though you may not have the guts to fire them, it's unlikely that you'll be stupid enough to promote them. Right? Let's hope so.

It takes a strong manager to deal with a brilliant jerk onboard. Expect her to fight you tooth and nail on all feedback. This isn't going to be easy on either of you. The difficulty is, if she doesn't see her behavior as a problem, she won't change it. It's unlikely that you alone will be able to convince her that her behavior is a problem. All the evidence in the world can't change a person who doesn't want to change.

The best thing you can do for your team, in the context of having a brilliant jerk, is to simply and openly refuse to tolerate bad behavior. This may be one of the few instances where "praise in public, criticize in private" is upended. When a person is behaving badly in a way that is having a visible impact on the team, and a way you don't want your culture to mimic, you need to say something in the moment to make the standard clear. "Please do not speak to people that way; it is disrespectful." You'll want to have tight control of your own reaction because delivering this in public is walking a fine line. If you seem emotional, it may undermine you. The offender may write off your feedback as just emotion, or you may come off as picking on the person. Keep the feedback neutral but to the point if you are going to deliver it in the moment, in public. Note that this approach should only be used for behavior you feel is detrimental to the whole group. If you just think the person is undermining you personally, discuss it in private. Your first goal is to protect your team as a whole, the second is to protect each individual on the team, and your last priority is protecting yourself.

THE NONCOMMUNICATOR

Another very common problem team member is the noncommunicator—the person who hides information from you, from his teammates, from his product manager. The person who prefers to work in secret and unveil a magical project when everything is done and perfect. The person who, instead of talking things out with teammates, reverts their commits, or takes their tickets and does the work for them. The person who doesn't want to go through code review and who doesn't ask for design review on big projects.

This team member annoys everyone around him. As the noncommunicator's manager, you have to nip this information-hiding habit in the bud as soon as possible. If necessary, make it clear that he's not meeting expectations for his work. This is often a sign of fear—the person is afraid that he'll be found lacking, or he'll be asked to do work he's not interested in. Sometimes it's a sign of a person who feels he should have more responsibility and who doesn't respect his manager. Whatever the cause, this person disrupts team cohesion because he isn't being collaborative with the rest of his teammates; he doesn't feel safe sharing his work in progress, and his fear often sets an example for the rest of the team.

If possible, address the root cause of the hiding. If the hider is afraid of being criticized, does your team have a harsh culture that needs to be addressed? Does your team have that psychological safety in general? Is the rest of the team treating this person like an outsider, perhaps because he has a different background or skill set? If the team is rejecting the individual, you will need to decide whether to try to correct the team or move the individual to another team. Sometimes, moving the individual is the kindest thing to do; other times, the best solution is to work with the team as a whole to change the balance of culture and break the habits that exclude new people.

THE EMPLOYEE WHO LACKS RESPECT

The third type of toxic individual is the person who simply doesn't respect you as a manager, or who doesn't respect her teammates. Addressing this person will be difficult and you may require some help from your manager, but if you can handle this yourself, it's a sign of great character. Simply put, if your team member doesn't respect you or her peers, why is she working there? Ask her if she wants to be working on your team. If she says she does, lay out what you expect, clearly and calmly. If she says she doesn't, start the process to move her to another team, or help her leave the company.

That's it? That's it. You can't have a person working for you who doesn't respect you, or doesn't respect your team. It will eat away at the cohesion of the rest of the team as they wonder whether that person is right in not respecting you. The sooner you pull off the Band-aid, the better.

Advanced Project Management

As an engineering manager, you will help set the schedule for your team. As the larger organization tries to figure out what the plans for the quarter or year might look like, you'll estimate whether your team can do certain projects, how much work those projects will be, and whether you have the right people to complete the work. You might be asked if your team can take on the support of old systems in addition to their current commitments, or how many people you'd need to hire in order to support a new initiative. The organization will expect you to be capable of doing both off-the-cuff estimation and more concrete project planning.

We gave a high-level overview of project management in Chapter 3's discussion of being a tech lead, but now I want to dig into some of the advanced work. As the manager of a team, while you may push some of the project planning onto your tech leads, you'll likely need to do some of that work yourself. You may have to decide which projects to take on, and when to push back on accepting projects. You'll probably be asked for rough estimates as to when work will be done, even work that is planned and iterated on in an agile fashion.

You need to have a strong sense of the rhythms and pace of your team to manage their workload successfully, but fortunately there are some shortcuts that can help.

PROJECT MANAGEMENT RULES OF THUMB

Here are some rules of thumb to keep in mind.

None of this is a replacement for agile project management

Before I begin, I want to make clear that I'm not suggesting that you go into waterfall mode and plan every project in detail from the get-go. However, most teams have both high-level, long-term goals, and short-term objectives that will enable them to meet those goals. When it comes to actually planning the details of the smaller pieces, an agile process where the team collaborates to divide and roughly estimate work is very effective at smoothing and organizing the day-to-day. As the manager, you're not trying to disrupt or even own that part of the execution process. However, you are responsible for the larger picture—the accomplishments that are measured in months instead of weeks—and this is where you have to start exerting some higher-level planning.

You have 10 productive engineering weeks per engineer per quarter

There are 52 weeks in a year, or about 13 per quarter. However, realistically your team will lose a lot of that time. Vacations, meetings, review season, production outages, onboarding new employees—all of these things take away from focus. Don't expect to get more than 10 weeks' worth of focused effort on the main projects per team member per quarter. It's likely that Q1 (immediately after the winter holidays) will be the most productive and Q4 (the quarter that includes winter and the end-of-year holidays) will be the least productive.

Budget 20% of time for generic sustaining engineering work across the board

By "generic sustaining engineering work," I mean testing, debugging, cleaning up legacy code, migrating language or platform versions, and doing other work that has to happen. If you make this a habit, you can use it to tackle some of the midsize legacy code every quarter and get decent improvements. Cleaning up systems as you go keeps those systems easy to work in, which keeps your teams moving forward on new features. In the worst case, you can use this slack to smooth over unexpected delays in feature development, but if you fill the schedule to 100% with feature development, expect that the feature development will quickly slow down as a result of this overscheduling.

As you approach deadlines, it is your job to say no

You will almost certainly have occasional deadlines, either goal dates that you've set or goal dates that came down from on high. The only way to achieve these goals is to cut scope at the end of the project. That means that you, as the engineering team lead, will partner with your tech lead and the product lead/business representative to figure out what "must-haves" are not actually must-haves. You will have to say no to both sides. There will be times when the engineering team will say that they can't possibly implement a feature without doing some other technical work, and you will need to figure out when to push for a hack implementation and when to hold back for the right implementation. There will be product features that require significant engineering complexity to implement, and you'll need to work with the product team to figure out the real must-haves while explaining the cost to get to their vision. When push comes to shove, you'll be the person to give the team options as to what can realistically be implemented, or how much more time getting everything in will require.

Use the doubling rule for quick estimates, but push for planning time to estimate longer tasks

The popular doubling rule of software estimation is, "Whenever asked for an estimate, take your guess and double it." This rule is appropriate and good to use when you're asked for an off-the-cuff guess. However, when you're talking about projects that you think will take longer than a couple of weeks, go ahead and double the estimate, but make it clear that you'll need some planning time before you're sure about the timescale. Sometimes the longer tasks will take far more than twice your estimate, and it's worth spending some time planning more carefully before you commit your team to a big, unknown project.

Be selective about what you bring to the team to estimate

Part of the reason that I stress your role in this estimation and planning process is that it's distracting and stressful for engineers to have a manager who's constantly asking them for random project estimates. As the manager, you're responsible for handling uncertainty and limiting how much of that uncertainty you expose to your team. Don't be a telephone between the engineers and the rest of the company, parroting messages back and forth and distracting people who are busy with the important tasks you've already committed to do. But you're not a black hole, either. Try to get a teamwide process in place for talking about new features and customer complaints, and limit estimations that occur outside of this process.

Ask the CTO: Joining a Small Team

I'm a newly hired manager for a team of five engineers. I have been a manager before at other companies, but I'm brand new to the company, the technology, and the team. How should I think about my time in these first few weeks?

Joining a small team as a manager is tough. It's one thing to balance technical work when you've been promoted to manager from software engineer, but it's another thing to come in new with a team to manage and new code to learn.

There are a few ways to get into the software without annoying the team. First, get someone to walk you through the systems and architecture, as well as the process for testing and releasing the software. If there is a normal developer onboarding process where you learn how to check out code and deploy the systems, go through that process. Spend some time getting comfortable in the code bases, and start watching the code reviews or pull requests, if they exist.

Plan to work on at least a couple of features in your first 60 days. Take a specced-out feature and add it. Pair with one of the engineers on a feature he's working on, and have him pair with you as you start working on a feature of your own. Get your code reviewed by a member of the team. Perform a release, and do a rotation of supporting the systems for at least a couple of days if support is part of the team's responsibilities.

You can probably tell that this means your management onboarding may go more slowly because you're also learning how to work in the systems. This slowdown is worth it. By getting to know the code, the processes for writing code, and the tools and systems your team use for their day-to-day, you will gain the understanding necessary for managing the team, and the technical credibility necessary for them to see you as a capable leader.

Assessing Your Own Experience

- What are your new responsibilities now that you're the manager of a team? What tasks have you stopped doing or handed off to someone else in order to make time for these new responsibilities?

- How well do you feel you know the day-to-day challenges of writing, deploying, and supporting code on your team?

- How often does your team mark work as completed?

- When was the last time you wrote a feature, debugged a problem, or paired with a member of your team on some code he or she was struggling with?

- Are there one or two team members who cause the bulk of negativity on the team? What is your plan for getting rid of the problem moving forward?

- Do your team members seem engaged with one another? Do they smile in meetings? Make jokes in chat? Get coffee or lunch together? When was the last time you all sat down together without talking about work?

- How does your team make decisions? Do you have a process for assigning decision-making responsibility? What decisions do you hold yourself responsible for making?

- When was the last time you reviewed a completed project to see if it had achieved its goals?

- How well does your team understand why they are working on the projects they are working on?

- When was the last time you cut scope on a project? What did you use to determine which pieces to cut?

6

Managing Multiple Teams

Welcome to the world of multiple-team management! We're going to talk about managing multiple teams before we talk about managing managers, because while those things are related, they don't necessarily coincide. You probably have tech leads reporting to you now, though, and juggling the work of directly managing more than three or four people with the process of understanding details about what's happening across a couple of teams probably means one important thing: you're not writing (much, any, production) code.

When I created the career ladder for my previous job, the director of engineering role was usually the place where a person would start to manage multiple large teams. Let's review some of the description from my engineering ladder:

> *The engineering director is responsible for a significant area of the technology team. The engineering director typically leads engineers across multiple product areas, or multiple technology functions. Both tech leads and individual contributors report into them.*
>
> *The engineering director is not generally expected to write code on a day-to-day basis. However, the engineering director is responsible for their organization's overall technical competence, guiding and growing that competence in the whole team as necessary via training and hiring. They should have a strong technical background and spend some of their time researching new technologies and staying abreast of trends in the tech industry. They will be expected to help debug and triage critical systems, and should understand the systems they oversee well enough to perform code reviews and help research problems as needed. They*

should contribute to architecture and design efforts primarily by serving as the technically savvy voice that asks business and product questions of the engineers on their teams, ensuring that the code we are writing matches the product and business needs and can scale appropriately as those needs grow.

The engineering director is primarily concerned with ensuring smooth execution of complex deliverables. To that end, they focus on ensuring that we continually evaluate and refine our development/infrastructure standards and processes to create technology that will deliver sustained value to the business. They are responsible for creating high-performance, high-velocity organizations, measuring and iterating on processes as we grow and evolve as a business. They are the leaders for recruiting, head-count management and planning, career growth and training for the organization. As necessary, directors will manage vendor relationships and participate in the budgeting process.

The impact of an engineering director should reach across multiple areas of the technology organization. They are responsible for creating and growing the next generation of leadership and management talent in the organization, and helping that talent learn how to balance technical and people leadership and management. They are obsessed with creating high-functioning, engaged, and motivated organizations, and they are expected to own retention goals within their organization. Additionally, engineering directors are responsible for strategically balancing immediate and long-term product-/business-focused work with technical debt and strategic technical development.

Directors are strong leaders who set the example for cross-functional collaboration both between technology and other areas of the company, and across divisions of technology. The goal of this collaboration is to create both a strategic and tactical tech roadmap that tackles business needs, efficiencies and revenue, and fundamental technology innovation. The director is a very strong communicator who can both simplify technical concepts to explain them to nontechnical partners and explain business direction to the technology team in a way that inspires and guides them. Directors of engineering help to create a positive public presence for Rent the Runway tech and are capable of selling the company and their area to potential candidates.

Due to their breadth of exposure to both technology and the business drivers, directors are responsible for guiding the goal-setting process for all of the teams in their organization, helping these teams articulate goals that support both business initiatives and technology and organizational quality.

I took pains to make sure that we called out the fact that engineering directors would not necessarily be writing code every day, because I believe that it is very difficult for a person responsible for hands-on management of multiple teams to write code. Your schedule, by this point, has probably moved away from "maker" and firmly into "manager." Between your 1-1s, meetings with other engineering leads, team planning sessions, and sessions with your peers in product management or other business functions, you're probably quite busy. Be realistic about your schedule at this point. If you don't have solid blocks of time to dedicate to it and you can't realistically guarantee solid blocks of time at least a few days a week, any code you write is going to be very slow-going.

Fortunately for us, there are ways to stay hands-on that don't require writing a lot of production code. Code reviews are a good thing to stay in practice with, at least as a secondary reviewer. If you created systems when you were more hands-on, stay engaged with those systems, because you'll remember the details better than most, and you can help engineers working in those systems with code reviews and questions. Debugging and production support are also valuable. How you stay hands-on depends on your skill set. If you were not a strong debugger before you went into management, jumping into incidents may be more annoying than useful. You may be more helpful doing pair programming, or fixing minor bugs or features. So often we diminish these small efforts as not worthwhile, but they're very good at keeping you in tune with the feeling of software development and showing your teams that you are willing and able to help out with the day-to-day in a valuable way.

The risk of going hands-off is greatly amplified if you don't spend enough time coding before moving into this role to get yourself deeply, fluently comfortable with at least one programming language. I advocate strongly that you spend the time to gain mastery of programming before moving into management. For me, this took about 10 years, including my undergraduate and graduate degrees. You may do it faster than I did, but scrutinize yourself carefully in this regard. Do you feel fluent enough in at least one programming language to productively contribute to a good code base written in it, after a limited time spent getting up to speed in the basics of writing it, using a standard development environment, and working in standard frameworks and libraries? Eventually even the deepest knowledge will atrophy, but fluency in working in a language (which includes comfort with its standard tools, libraries, and runtimes) is something that sticks with you for a long time.

Useful fluency also requires an ingrained understanding of what it means to work productively in such a language, hopefully on a team with other people building production software. Without this sense of the rhythms of building software, you'll struggle with one of the critical parts of the job at this level: debugging team issues and keeping your teams producing quality software smoothly.

Finally, even if you don't intend to write much code, I strongly advise you to keep at least a solid half-day once a week completely free from meetings or other obligations, and try to use this time at least partially on some creative pursuit. You might write blog posts for your engineering blog, prepare conference talks, or participate in an open source project. Do something to scratch that creative itch, which can otherwise be hard for you to scratch as a manager.

Ask the CTO: I Miss Code!

I'm managing two complex teams and my management responsibilities are forcing me to step back from technical responsibilities. I find that I miss code terribly. Is this a sign that I shouldn't be a manager?

Almost everyone who goes from a heavily hands-on technical role into management has a transition period where they question frequently whether they've made a mistake. Furthermore, many worry that they're losing all of their valuable skills in the process. Ask yourself whether you've internalized the idea that management is not a job. The tech industry is filled with people who despise management, thinking it's not as important a job as writing code. But management is a job, it is a necessary and important job, and in particular, it's your job right now.

Writing code is full of quick wins, especially for the experienced developer. You make tests pass, you see new features come to life, you get something to compile, you fix a problem. Management has fewer obvious quick wins, especially for new managers. It's natural to feel some longing for simpler times, when it was just you and your computer and you didn't have to deal with all these messy, complicated humans. You probably felt some similar nostalgia for your school days when you first started working full-time, because by the time you left school you knew exactly what to expect from it. It's OK to feel nostalgia for the simpler times, and a little bit of fear for what you're giving up. But you can't do everything all at once. Becoming a great manager requires you to focus on the skills of management, and that requires giving up some of your technical focus. It's a tradeoff, and one you'll have to decide if you're up to making.

Managing Your Time: What's Important, Anyway?

When you have so many management duties that you have little time to code, you can start to feel like your day has been taken hostage by the whims of others. You start to see the meetings pile up: the one-on-ones, the planning meetings, the status updates. Standups. Sit-downs. Fight fight fight!

Wait, no—no fighting in the war room!

It's time—now—for you to figure out how to manage your time. Otherwise, you'll find yourself with days gone by and little to show for them. You still have responsibilities as a manager. You still have deliverables that require you to do more than sit in a meeting—things like setting goals for the team, helping your product team put details on the product roadmaps, and making sure that an assigned task actually got finished. That last one, following up on task completion, can become one of the biggest time sinks and distractions in your day if you aren't careful.

Time management is a personal thing. Some people are very organized, and those people develop complex strategies for managing their calendars and to-do lists. I applaud these people. I am not generally one of them. However, I have found the ideas in David Allen's book *Getting Things Done*[1] to be useful to think about, and I recommend reading it even if you don't adopt the whole process.

In the meantime, my general time-management philosophy will serve you well no matter what your tactics might be. Managing your time comes down to one important thing: understanding the difference between *importance* and *urgency*. Pretty much all of your tasks will fall on a graph of these two elements. Roughly, they are in one of four quadrants (see Table 6-1).

---

1 David Allen, *Getting Things Done: The Art of Stress-Free Productivity* (New York: Penguin, 2001).

Table 6-1. Prioritizing your time

	Not urgent	Urgent
Important	Strategic: make time	Obvious work
Unimportant	Obvious avoid	Tempting distractions

If it's important and urgent, you're doing it. You know what I'm talking about. There's a major outage that you're helping to fix. Performance review write-ups are due tomorrow. You want to make an offer to a great candidate who has another competing offer expiring in two days. If you drop the ball on a task in this category, you lose something tangible. It's unlikely that you're blind to the need to get these sorts of things done.

A big part of the challenge of time management emerges when you start to lose the sense of importance. Urgency is often more clearly felt than importance. Responding to email is a good example. It's easy to get sucked into email as a distraction, because the red dot tells you there's something new, and it feels urgent to you to acknowledge it. And yet how often is email really urgent? Email is probably the worst vehicle for conveying urgent, time-sensitive information. It feels urgent, but it isn't urgent. This is why so many precise time-management tips encourage reading and responding to email at specific times of the day. We also tend to substitute *obvious* for *urgent* in determining something's value. If a meeting is on your calendar, it's obvious where you should be at that time, but is that meeting really urgent, or are you using it to avoid thinking about the best way to use your time?

There are many things that feel urgent that aren't. The whole of the internet, for example. News, Facebook, Twitter. Chat can feel urgent, but chat is almost as bad as email for communicating truly urgent and important information in the case of a collocated team. We've moved a lot of communication in the modern tech workplace out of email into chat systems like Slack and HipChat. This has benefits and drawbacks, but it's important to note that moving communication is not the same thing as eliminating communication. The words and information keep flowing, they just move to different places, and you can get even more distracted by the ever-moving trickle of information in chat.

It's likely that you're spending a lot of your time on things that are urgent but only slightly important, and sacrificing things that are important but not urgent. One example of an important but not urgent task is actually preparing for meetings so that you can guide them in a healthy way. Healthy meetings require involvement from all parties, and a culture that favors short but productive meetings requires that participants do some up-front work to come to the meeting prepared. As a manager of multiple teams, you can win back a lot of time by pushing an efficient meetings culture down to your teams. Hold people accountable to prepare in whatever way makes sense. Ask for agenda items up front. Any sort of standard meeting that involves a group of people, whether it's planning, retrospective, or postmortem, should have a clear procedure and expected outcomes.

One of the major changes at this level compared to the previous level is that your boss will expect you to be mature enough to manage yourself and your teams independently. This means that your manager trusts you to proactively deal with all those important but not urgent things before they become urgent, and especially before they become urgent for your manager. No one will tell you how to manage your calendar to give yourself the time to do this. I've seen managers fail at this point because they just could not juggle all the different tasks in an organized fashion.

Meetings can fall into that urgent but not important category, and you may decide to simply not attend them where you're not clearly needed. Be very careful with over-deploying this strategy at this particular level of management. The responsibility of keeping your teams successfully moving forward and happily engaged is on your shoulders. When you stop going to all of their internal meetings, you run the risk of missing out on the very clues that will help you catch problems early—a major one being the existence of too many boring meetings. During meetings, look around the room at your team and notice their engagement. If half of them are falling asleep, staring off into space, on their phones or laptops ignoring the proceedings, or otherwise disengaged, the meeting is wasting their time. Your attendance at these meetings is partially to pay attention to the dynamics and morale of your team. A happy team will feel energized and engaged. An unhappy or unmotivated team will feel drained or bored.

Back to matters important but not urgent. Thinking about the future is high on this list. There are undoubtedly things that you know you should do but have put off. Perhaps it's writing job descriptions for roles you're hiring for. Perhaps it's developing a hiring plan at all. It might be reviewing the current work on a project to make sure no obvious problems are creeping in, or talking to a manager on another team where there is conflict or a difference of opinion on how to move forward on a shared issue. It might be cultivating the list of things that are important but you haven't thought about in a while, so you know what to focus on. If you don't set aside some time to focus on these issues, they'll sneak up on you in negative ways. As a manager of multiple teams, you're responsible for balancing breadth and depth of thinking, for knowing the details of your teams today but also looking at where you need to be going in the future and what you need to get there.

As you navigate your new obligations, start to ask yourself: How important is the thing I'm doing? Does it seem important because it is urgent? How much time have I spent this week on urgent things? Have I managed to carve out enough time for things that are not urgent?

The Hardest, Shortest Lesson of Becoming a Manager

As a manager, I have this mental list of things about what my team needs. Things that I'm monitoring, things that I'm trying to fix, things that I'm trying to find for them. It's my job to understand what is going on and what the team as a whole needs to be effective.

Maybe you can look at the state of things and say, "We have a deadline right now, and what we need is another engineer for the next month. That engineer is me."

But more likely you look at the state of things and realize that what your team needs is a manager. Because you need to hire X more people. Because Y has a lot of potential but needs some coaching. Because product or design or some other team hasn't given you what you need, so you need to go and get it. Because process is important, and the process you have is insufficient or just plain wrong.

If your team needs a manager more than they need an engineer, you have to accept that being that manager means that you by definition can't be that engineer. I know some people manage both, but you need to decide, if you're going to suck at one, which one that will be.

I feel bad when I suck at being an engineer, but sucking at being a manager would be a choice I inflicted on other people. That's not fair.

So at the end of another day when I feel like I didn't write enough code and I have no way to quantify what I've achieved, I tell myself I was being as good a manager as I know how to be. And that has to be enough for today.

*—Cate Huston (http://bit.ly/
huston-manager)*

Decisions and Delegation

How do you feel at the end of the day these days? If you're like many new full-time managers, you probably feel quite drained. Even though you didn't write a lot of code—or any code!—all day, when you get home you find yourself with no energy to decide what to eat for dinner, no energy for hobbies, and the desire to eat comfort food, drink a beer perhaps, and stare blankly at the computer or TV until it's time to go to bed.

The first several months of managing multiple teams can feel like a death march, even when your hours are not excessive. Your once-focused attention gets sliced and diced between the various meetings that pepper your day. I lost my voice repeatedly during my first few months managing multiple teams; I was totally unused to talking so much every single day. A friend of mine recently became a director of engineering, and she had to start having an assistant order her lunch because she discovered that she would forget to eat—and had no energy to decide what to eat when she realized she needed food.

So, first, the bad news: the only way out of this situation is to go through it. In fact, I would expect most people to go through this for a while. If you haven't experienced it at all, either count yourself extremely lucky, or double-check to make sure you're really paying attention to everything that needs your attention. In my experience both going through this transition and managing people in it, if you don't feel a little bit overwhelmed, you're likely missing something.

The best way to describe the feeling of management from here on out is plate spinning. If you're not familiar with it, plate spinning is a fancy form of juggling where the juggler has several poles, each with a plate spinning on top of it. The juggler must attend to each plate before it slows down enough to fall off the pole. Your plates are the people and projects you're overseeing, and your job is to figure out how much attention each one needs at what time. It's important that you approach this spinning with a student's mind. You're still learning how to spin plates, and you're going to drop some on the floor because you've neglected them for too long. Honing your instincts about when to touch which plate is the name of the game.

Now for the good news: you'll get better at this over time. Your instincts will improve. You'll start to recognize the early warning signs of projects that are going poorly, people who are getting ready to quit, and teams that are underperforming. I recommended in the last section that you think carefully about dropping out of meetings, and part of the reason is that those meetings are where you learn what healthy and unhealthy dynamics look like. This is also why I strongly advise you maintain your practice of regular, reliable 1-1 meetings with everyone who reports directly to you. If you have too many people, you may need to shorten those meetings or hold them biweekly instead of weekly, but skipping 1-1s because you're too busy with other things is a great way to miss the warning signs of an employee who is going to quit.

I called this section "Decisions and Delegation"—so where does delegation fit in? Delegation is the primary way you claw yourself out of the feeling of having too many plates spinning at once. As tasks come at you, ask yourself: do I need to be the person who completes this work? The answer may depend on a few factors (see Table 6-2).

Table 6-2. Deciding when to delegate or do it yourself

	Frequent	Infrequent
Simple	Delegate	Do it yourself
Complex	Delegate (carefully)	Delegate for training purposes

The degree of complexity and the frequency of the task can act as guides to determining whether and how you should delegate.

DELEGATE SIMPLE AND FREQUENT TASKS

If the task is simple and frequent, find someone to whom you can hand it off. Examples of simple and frequent tasks might include running daily standups, writing up a summary of the teams' progress each week, or conducting minor code reviews. Your tech leads or other senior engineers can take responsibility for these tasks and probably won't even need training to do so.

HANDLE SIMPLE AND INFREQUENT TASKS YOURSELF

If it's faster to do something yourself than it would be to explain it to someone else and it rarely needs to be done, roll up your sleeves and do it, even if you deem the task beneath you. This can be anything from booking the occasional conference ticket for your team to running the script that generates quarterly reports.

USE COMPLEX AND INFREQUENT TASKS AS TRAINING OPPORTUNITIES FOR RISING LEADERS

Tasks like writing performance reviews and making hiring plans are yours alone. However, these are also the skills you will want to pass on to rising managers. You may have a tech lead sit with you to write the performance review for an intern, or have a senior engineer provide feedback on how many new people he believes would be needed to support a project next year. Ask for help from above on these tasks until you feel comfortable doing them, but once you feel comfortable, start pulling in rising leaders to learn how they are done.

DELEGATE COMPLEX AND FREQUENT TASKS TO DEVELOP YOUR TEAM

Tasks like project planning, systems design, or being the key person during an outage are the biggest opportunity you have to grow talent on your team while also making the team run better. Strong managers spend a lot of their time developing members of their teams in these areas. Your goal is to make your teams capable of operating at a high level without much input from you, and that means they'll need individuals who can take over these complex tasks and run them without you around.

Are your teams learning how to operate independently, or are you keeping them dependent on you for critical functions? List the tasks that you and only you really know how to do for the team. Some of them may be appropriate, like writing performance reviews or making hiring plans, but many of them are important to teach your teams to accomplish themselves. Project management. Onboarding new team members. Working with the product team to break down product roadmap goals into technical deliverables. Production support. These are all skills members of your team need to learn. Teaching them may take time up front, but in the long run it will save you time. Not only that, but teaching your team how to do these things is part of your job. It is your responsibility, as a manager, to build up the talent in your organization, and to help your people learn new skills they'll need for the next stages of their careers.

Delegation is a process that starts slow but turns into the essential element for career growth. If you teams can't operate well without you around, you'll find it hard to be promoted. Develop your talent and push decisions down to that talent so that you can find new and interesting plates to learn how to spin.

Ask the CTO: Warning Signs

I've experienced a couple of times now when teams have struggled unexpectedly, and a person has quit without warning. Are there any warning signs I can look out for to catch these issues earlier?

There are definitely signals that you start to notice after you've been managing for a while. Here are some I've learned to spot:

- *The person who is usually chatty, happy, and engaged suddenly starts leaving early, coming in late, taking breaks to leave during the workday, staying quiet in meetings, and not hanging out on chat.* This person is either having a major personal issue or getting ready to quit. Usually, people will tell someone when there is a personal issue (such as a sick relative, relationship problems, or health issues), but not always. If this happened right after a major adjustment, such as a promotion, a team reorganization, or other event, the person may feel that she was overlooked. Regardless of the reason, you may want to have an honest conversation and try to get to the root of the issue before she resigns.

- *The tech lead claims that everything is going well, but skips your 1-1s frequently and rarely provides detail in his status updates.* This person may be hiding something. Often what he's hiding is that the progress is going far slower than he anticipated, or he's building something outside of the scope of the project. Help him create a clear project plan early and set expectations for how to adjust that plan when things change, so that it's harder for him to hide a lack of progress. Also help him clarify the project's goals and scope, which can be daunting for some new tech leads. You may have experienced something similar when managing new hires who were in over their heads. This is also related to the person who spends a lot of time advocating for new languages/platforms/processes instead of finishing her work.

- *The team has absolutely no energy at all in their meetings. In fact, the meetings feel like a total slog, with the product manager and tech lead doing all of the talking while the rest of the team sits silently or speaks only when called upon.* A lack of engagement in meetings tends to mean the team isn't engaged by the work or do not feel like they have a say in the decision-making process.

- *The team's project list seems to change every week, depending on the customers' whims that day.* This team hasn't thought about its goals beyond pleasing customers, and may need better product or business direction.

- *A small team internally seems very fragmented in understanding; the engineers profess ignorance about systems they don't work on and lack the curiosity or openness to learn about those systems.* This team is more strongly identified by their day-to-day work and the systems they touch than by the larger team or the company. They may be resistant to changing their systems based on the needs of the larger team or the business.

Challenging Situations: Strategies for Saying No

A manager's job involves making it easy for her employees to get things done by creating fertile environments in which work can happen. She focuses her team so that they can do what they do best. She cultivates camaraderie and friendship on the team, and helps people learn new skills. In all these things, she is an enabler, a coach, and a champion.

But to create this environment, she sometimes must say no. She must say no to the team. She must say no to her peers. She must even say no to her boss. Each of these nos is hard, in its own way, and a strong manager must develop effective strategies for saying no. Here are a few that I have identified.

"YES, AND"

Saying no to your boss rarely looks like a simple "no" when you're a manager. Instead, it looks like the "yes, and" technique of improvisational comedy. "Yes, we can do that project, and all we will need to do is delay the start of this other project that is currently on the roadmap." Responding with positivity while still articulating the boundaries of reality will get you into the major leagues of senior leadership. This type of positive no is a pretty hard skill for most engineers to master. We're used to articulating the downsides of projects, and getting out of the knee-jerk "no, that can't happen" habit is hard. Start mastering the "yes, and" strategy for saying no, particularly when interacting with your boss and peers, and see how it often transforms contentious disagreements into realistic negotiations for priority.

CREATE POLICIES

When it comes to your team, you want to help them understand what it takes to get to "yes." Perhaps you are dealing with an engineer who wants to switch to a new programming language for a project, one that your team doesn't use. He has some great arguments as to why this language is the perfect tool for the job, but you're reluctant to add a new tool just because it's perfect. You might be tempted to just say no, give the reasoning, and leave it, and sometimes that will work. But you may find yourself saying the same "no" over and over again, giving the same reasons. "No, we need to have more people who know that language; we need to understand what it means to put that language into production." "No, we need to have standards for logging; we need to think about what testing would look like." When you start repeating yourself, you have the basis for a reasonable policy. That policy consists of the hard requirements that must be met in order to say yes, and some guidelines for thinking about the decision. Making a policy helps your team know in advance the cost of getting to "yes."

"HELP ME SAY YES"

Policies are useful, but they don't cover every case. The next strategy, "help me say yes," is similar to policy writing, but works better for the one-off instances where there is no clear policy. Sometimes you'll hear ideas that seem very ill-considered. "Help me say yes" means you ask questions and dig in on the elements that seem so questionable to you. Often, this line of questioning helps people come to the realization themselves that their plan isn't a good idea, but sometimes they'll surprise you with their line of thinking. Either way, curious interrogation of ideas can help you say no and teach at the same time.

APPEAL TO BUDGET

When it comes to both your team and your peers, one tactic that you can use is appealing to time and budget. Lay out the current workload in plain terms, and show how there is little room to maneuver. Sometimes this is coupled with "not right now," another somewhat passive-aggressive way of saying no. "Not right now" implies that you might agree with the idea but can't do it at this moment, so maybe you'll get to it in the future. This is frequently true, and so it's easy to fall back into "not right now" mode. But as I discussed earlier, when you give an implied promise that "not right now" means that you'll seriously do something "later," you need to be sure that later can actually happen.

WORK AS A TEAM

Speaking of your peers, there will be times when you and your peers (particularly across functions—i.e., your product or business peers) will need to act together to say no. This can apply to a no at any level. Sometimes you will use your technical authority to say no in a way that is beneficial to the product team. Sometimes you will appeal to the finance department to help you say no to certain budget excesses. Playing good cop/bad cop can be a little bit dishonest, so use this sparingly, but it can be useful to lend your authority to a no and then be able to call in the favor when you need support for your own no in the future.

DON'T PREVARICATE

When you know that you need to say no, it's better to say it quickly than to delay and drag out the process. If you have the authority to say no, and you don't believe something should happen, do yourself a favor and don't agonize over the process. You'll be wrong sometimes, so when you discover that you were too quick to say no, apologize for making that mistake. You won't have the luxury to carefully investigate and analyze every decision, so practice getting comfortable with the quick no (and the quick yes!) for low-risk, low-impact decisions.

Ask the CTO: My Tech Lead Isn't Managing

I have a tech lead who was supposed to be overseeing one of our junior engineers on a project to rewrite our app from Objective-C to Swift. I just found out that the junior engineer still hasn't created a project plan, and hasn't answered any of the feedback I gave in the design review. How do I get the tech lead to manage this without me having to step in?

Delegation failures happen. It sounds like your tech lead doesn't understand that you're holding her responsible for making sure that the junior engineer follows up on design feedback and creates a project plan. So the first step is to ask the tech lead why these things haven't happened yet.

The answer you will get is likely to be a combination of things. One, the tech lead is busy with her own work and didn't remember to follow up with the junior engineer. This happens, and you'll have to remind her that mentoring and overseeing this person's work needs to be scheduled in with her code and other responsibilities.

Two, the tech lead may not know how to push the junior engineer when he doesn't want to commit to a schedule. Ask her how she's tried to get the information from him, and see whether you have an opportunity to suggest different approaches. Sometimes new tech leads are reluctant to push people for project plans because they don't feel that they have authority and are flustered when they ask for something and the other person just never delivers.

The best thing to do here is to work with your tech lead to give her the skills and confidence to ask for reports from other members of the team. It will be slower than stepping in and asking for them yourself, but you'll teach the team to respect her requests and teach her how to lead the team independently.

Technical Elements Beyond Code

Management at this level gets confusing. We hire managers based partially on their technical skills, but many of us think of this job as not really "technical." After all, the manager is probably not writing much code or doing much systems design, right?

Assuming that the job at this level becomes essentially nontechnical is a mistake. It turns out there's more to running effective engineering teams than pure management skills, and management at this level will require you to learn some new skills that are easiest to learn if you understand the practice and discipline of software engineering. You're going to turn your technical focus now to observing and improving the systems of work that your developers are operating within. In particular, you now need to develop an eye for the technical health signals for the overall team. But what are those health signals?

The popular management book *First, Break All the Rules*[2] discusses several questions you can answer to help predict team productivity and satisfaction. Among them are:

- Do I know what is expected of me at work?

- Do I have the materials and equipment I need to do my work right?

- Do I have the opportunity to do what I do best every day?

For most engineers, the answer to these questions can be discerned by the speed and frequency with which they push code. If the work they need to do is clear, they know what code to write. If the tools, tickets, automation, and process are available and easy to use, they are able to get the code written. And if they're

---

2 Marcus Buckingham and Curt Coffman, *First, Break All the Rules: What the World's Greatest Managers Do Differently* (New York: Simon & Schuster, 1999).

not distracted by excessive meetings or drowning in support and incident management, they'll get a chance to write code every day. These health signals—frequency of code releases, frequency of code check-ins, and infrequency of incidents—are the key indicators of a team that knows what to do, has the tools to do it, and has the time to do it every day.

Measuring the Health of Your Development Team

When you're focusing on the health of your development teams, put on your technical hat to design systems and processes that will keep things moving. Create the tools that developers need to do their jobs. Help them focus so they can easily figure out what needs to happen next. Interrogate every process to determine the value it should provide, and always ask yourself if it can be automated further. Consider these ways to measure the health of your team.

FREQUENCY OF RELEASES

In Chapter 5 I talked about how a common dysfunction of technical teams is not shipping code, and release frequency is the most direct measure of this. If your company doesn't appreciate the value of releasing code frequently, I'm sorry. In this modern era, frequency of code change is one of the leading indicators of a healthy engineering team. Great engineering managers of product-focused teams know how to create environments where their teams can move fast, and part of moving fast requires breaking work down into small chunks. Even if your company doesn't appreciate it, you must work to help your team achieve the best release frequency possible for their product. Lest you think this doesn't apply to you because you're building a product (say, a database) that can't release frequently, I'm sure there is a complete artifact pushed to a beta/developer testing environment that provides a similarly valuable measure in terms of frequency and stability.

Why don't you release more frequently? Take a look at your team. If they don't release continuously, or daily, what does the release process look like? How long does it take? How often has something gone wrong in the past few months regarding releases? What does it look like when things go wrong? How often have you had to delay or roll back a release due to problems? What were the impacts of that delay or rollback? How do you determine if code is ready to go into production? How long does that take? Who is primarily responsible for determining this?

I bet if you honestly take a look at a team that isn't releasing frequently, you'll see cracks. The process of performing a release takes a long time. Engineers don't feel ownership over their code quality, and they leave all of that work to a QA team, which creates a lot of back-and-forth communication delays. Rolling back code in the case of a bad release takes a long time. Things go wrong in the process of releasing that lead to incidents in production (or broken development builds). A whole host of ills in a team come from not being able to release frequently.

Now, you might say, "Thanks for the advice, but I don't have the time to work on this with the product roadmap we have to deliver." Or, "Our systems aren't designed to be released frequently." Or, "It's just not that important that we change things that frequently."

Here's the thing. Is your team working to its full capacity? Are your engineers challenged and growing? Is your product team excited by the progress you're making? Are people able to spend most of their time on writing new code and evolving the systems? If so, great. Ignore me. You've got it under control. If not, you have a problem, and you're ignoring that problem at your peril.

It's important to remember that, as a technical leader, while you may not be writing code much, you're still responsible for the technical side of getting work done. You're also responsible for keeping your team happy and productive, and often the solution to this is not cheerleading or paying them better or praising them more, but instead enabling them to be more productive, challenging them to go faster and do better work, and helping them find the time they need to make their work more interesting. You have to be the advocate and push for technical process improvements that can lead to increased engineer productivity, even if you're not implementing them all yourself.

The beauty of pushing for more frequent releases is that it often uncovers a host of interesting challenges. There's no one true way to increase release frequency because frequency problems will be somewhat unique from team to team. You'll almost certainly need to solve some automation elements. Developer tooling to enable feature toggles that make sense for your code bases is another frequent challenge. Thinking about architecting code to move forward without breaking backward compatibility, rolling upgrades of systems, and implementing small changes instead of giant patches—all of these things may need addressing. You're responsible for leading the effort here, even if you don't do the work. Push for time away from the product roadmap to support increasing engineering productivity, and set goals for the team that inspire them to move faster.

FREQUENCY OF CODE CHECK-INS

It's hard to have an agile team that doesn't understand the value of breaking its work down into small chunks. You'll likely need to teach this skill to new hires out of college, but even senior developers sometimes need a push here. I'm not going to advocate for any particular software development methodology, but I find that engineers who don't write tests often have a harder time breaking down their work, and learning how to do test-driven development (even if they don't actually practice that on a daily basis) can help them get better at this skill.

I focus on this topic because it can be very uncomfortable for you as a new manager to tell people who may have been writing code for as long or longer than you have that their style needs updating. Conflict avoidance runs deep in most of us, and matters of what feels like personal style can be particularly difficult. If your company expects fast-moving product development, engineers who want to go off for weeks and write code alone without pushing it into shared version control will slow your team down and cause problems. You're not managing a research team. (Are you? Skip this section, then!) It's OK for you to have the expectation that work in progress is regularly updated.

FREQUENCY OF INCIDENTS

How stable is the software being produced by the team? Is the quality improving, getting worse, or staying the same? Determining the level of software quality you need for the product you're building and adjusting that measure over time is a technical challenge for you, the manager, to help address. If you're building a brand new product for a small but growing business, it may be more important to focus on features over stability. On the other hand, if you own mission-critical systems, stability and incident minimization may be your top priority. The goal here is to balance risk in such a way that neither incident frequency nor incident prevention turns into a job that takes developers away from writing code for days at a time.

You may work for a company that has developers support the code or systems they write. This process has some downsides; significantly, expecting members of a team to frequently be on-call nights and weekends is a huge contributor to burnout. Despite that risk, it has the upside of putting the best people to help fix a problem in the role of responding to it. As a manager, you may be tempted now to take yourself out of this role. I sympathize, but if your team is set up to do its own incident management, you should be moving yourself into the role of escalation support. You won't necessarily manage incidents as frequently, but you'll be expected to be available more often in case the person supporting the systems needs you.

Analysis around incident management should include the question, "Is our current setup enabling my team to do what they do best every day?" Incident management, when it becomes merely reacting to incidents rather than working to reduce them, can turn into a task that diminishes your team's ability to do what they do best. Engineers go on-call, they get burned out and exhausted from handling the deluge of problems and getting nothing done but fixing the consequences of incidents, and then they hand off the job to the next poor sap on the rotation. If that describes your team's approach to incident management and on-call, your team is not able to do what they do best every day, and every time they go on-call they probably hate their jobs a little bit more. In this case, as a leader you probably want to focus on providing time to actually design systems that are more stable, or writing code to fix the recurring incidents as they arise.

An overemphasis on incident prevention can also reduce your team's ability to do what they do best every day. Overfocusing on building systems that are defect-free, or pushing for error prevention by slowing down the development process, is often almost as bad as moving too fast and releasing unstable code. When risk reduction turns into weeks of manual QA, excessive and slow code reviews, infrequent releases, or a drawn-out planning process, the increased analysis can leave developers idle and restless, without necessarily reducing the risk of incidents.

Good Manager, Bad Manager: Us Versus Them, Team Player

Diana has just joined a midsize startup to run the long-neglected mobile team. She was told coming in that the team was a mess, so her first step is to quickly hire in a bunch of new people who worked for her at BigCo. They aren't quite a culture fit, and the team quickly turns into a clique of developers who see themselves as better than the rest of the organization. While the technology has improved, they seem to be clashing a lot with the product team, and ultimately the apps aren't evolving very quickly. After a year, Diana gets fed up with the company and quits. The rest of her new team soon follows, leaving the company right back where it started.

It can be hard for new managers to create a shared team identity. Many of them default to an identity built around the specifics of their job function or technology. They unite the team by emphasizing how this identity is special as compared to other teams. When they go too far, this identity is used to make the team feel superior to the rest of the company, and the team is more interested in its superiority than the company's goals. Rallying a team in this way is a *shallow binding* that is vulnerable to many dysfunctions:

- **Fragile to the loss of the leader.** In-group teams tend to be very fragile to the loss of their leader. When you hire a manager who builds a clique, that clique is likely to dissolve and leave the company if the manager leaves the company. This problem makes it that much harder to address the problems that the manager is causing by creating a clique in the first place.

- **Resistant to outside ideas.** In-groups tend to be resistant to ideas that do not come from those within the group. This means that they miss opportunities to learn and grow. The lack of growth for members of the team often causes the best members of the team to leave not just the group, but the company. Because they believe they're in the best group but they still find themselves bored, they don't appreciate the growth they could find just by switching to a new team.

- **Empire building.** Leaders who favor an us-versus-them style tend to be empire builders, seeking out opportunities to grow their teams and their mandates without concern for what is best for the overall organization. This often results in competition with other leaders for headcount and control of projects.

- **Inflexibility.** These groups tend to struggle against change that comes from outside the group. Reorganizations, cancelled projects, and shifting focus all can cause breaks in the core parts of their identity. Whether it's a move from functional groups to cross-functional teams, delaying the iPad application, or prioritizing a new product, the change can devastate the fragile bonds of the team to the company.

As a manager, be careful about focusing on your teams to the exclusion of the wider group. Even when you have been hired to fix a team, remember that the company has gotten this far because of some fundamental strengths. Before you try to change everything to fit your vision, take the time to understand the company's strengths and culture, and think about how you're going to create a team that works well with this culture, not against it. The trick is not to focus on what's broken, but to identify existing strengths and cultivate them.

Neil has also joined a startup where things are chaotic. While he can see that he needs to change the team, he moves cautiously to fire people and takes the time to make sure that new hires are always vetted by someone who's been with the company for a while. He spends a lot of time working closely with his peers in product, and proposes a path forward that emphasizes cross-functional collaboration. He focuses on setting clear goals and communicating them to his team. Things start out slow, but over time the entire organization feels stronger and both the technology and the product have improved dramatically.

Durable teams are built on a shared purpose that comes from the company itself, and they align themselves with the company's values (see "Applying Core Values" on page 200 in Chapter 9 for more on this topic). They have a clear understanding of the company's mission, and they see how their team fits into this mission. They can see that the mission requires many different types of teams, but all of the teams share a set of values. By creating a strong and enduring alignment between the team, its individuals, and the overall company, this *purpose-based binding* makes teams:

- **Resilient to loss of individuals.** While the clique is fragile, especially to the loss of the leader, the purpose-driven team tends to be very resilient to the loss of individuals and leadership. Because they're loyal to the mission of the larger organization, they can see a path forward even through loss.

- **Driven to find better ways to achieve their purpose.** Purpose-driven teams are more open to new ideas and value changes that can help them serve their purpose better. They care less about the source of an idea than its merit in achieving their goals. The members of these teams are interested in learning from others outside their function, and they actively seek out chances to collaborate more broadly to create the best results.

- **First-team focused.** Leaders who are strong team players understand that the people who report to them are not their first team. Instead, their first team is their peers across the company. This first-team focus helps them make decisions that consider the needs of the company as a whole before focusing on the needs of their team.

- **Open to changes that serve their purpose.** The collaborative leader understands that changes will happen to serve the wider purpose. Teams will change structure and people will need to move to where the business needs are. With that knowledge, these leaders create teams that are more flexible and understanding of frequent change in service of the larger vision.

Getting clarity about the purpose of your team and your company can take time. In startups especially, there is often some confusion about the current goals and even sometimes the underlying mission. In the case where the goals are fuzzy and the mission is unclear, do your best to understand the company culture and think about how you can set your teams up to work well within that culture. By collaborating across teams and across business functions, your teams will come to understand the bigger picture and appreciate their mission as part of that picture.

The Virtues of Laziness and Impatience

I love Larry Wall's idea that "laziness, impatience, and hubris" are virtues of engineers, as he articulated in *Programming Perl*.[3] These virtues persist into leadership, and learning how to channel these traits into advantages is something I encourage all managers to do.

As a manager, when you are dealing with people one-on-one you probably don't want to be impatient, of course. Impatience can be rude if it's directed at individuals. And you don't want to seem lazy, because there's nothing worse than working for a manager who seems to be taking it easy while you kill yourself to deliver projects. But impatience paired with laziness is wonderful when you direct it at processes and decisions. Impatience and laziness, applied to process, are the key elements to focus.

As you grow more into leadership positions, people will look to you for behavioral guidance. What you want to teach them is how to focus. To that end, there are two areas I encourage you to practice modeling, right now: figuring out what's important, and going home.

I can't stand watching people waste their energy approaching problems with brute force and spending time rather than thought. And yet, any culture where you are encouraged to work excessive hours all the time is almost certainly doing just that. What is the value in automation if you don't use it to make your job easier? We engineers automate so that we can focus on the fun stuff—and the fun stuff is the work that uses most of your brain, and it's not usually something you can do for hours and hours, day after day.

So be impatient to figure out the nut of what's important. As a leader, any time you see something being done that feels inefficient, question it: Why does

---

3 Tom Christiansen, brian d foy, Larry Wall, and Jon Orwant, *Programming Perl*, 4th edition (Sebastopol, CA: O'Reilly, 2012).

this feel inefficient to me? What is the value in the thing we are doing? Can we deliver that value faster? Can we strip down this project into something simpler and get it done more quickly?

The problem with this line of questioning is that often when managers ask whether something can be done faster, what they explicitly or implicitly want to know is whether the team can work harder or longer hours to deliver it in fewer days. This is why I encourage you to develop and show the value of laziness. Because "faster" is not about "the same number of hours but fewer total days." "Faster" is about "the same value to the company in less total time." If the team works 60 hours in a week to deliver something that otherwise would've taken a week and a half, they haven't worked faster, they've just given the company more of their free time.

This is where going home comes in. Go home! And stop emailing people at all hours of the night and all hours of the weekend! Forcing yourself to disengage is essential for your mental health, believe me. Burnout is a real problem in the American workforce these days, and almost everyone I know who has worked sustained excess hours has experienced it to some degree. It's terrible for individuals, terrible for their families, and terrible for teams. But this isn't just about preventing your own burnout—it's about preventing your team's burnout. When you work later than everyone else, when you send those emails at all hours, even if you don't expect your team to respond to those emails or work those hours, they see you doing it and think it's important. And that overwork makes them less effective, especially at the detailed knowledge work that engineers need to perform.

When you're a newish manager and you haven't figured out the tricks to do your job effectively, you might find yourself needing to work more hours to get it all done. That's OK for a little while. But I encourage you to figure out a way to work those hours without encouraging your team to do so, or making them feel obligated to be on your schedule. Queue up the weekend and overnight emails for the next workday. Put your chat status as "away" in off hours. Take a vacation, and don't answer email during that time. And constantly ask yourself the same questions you ask your team: Can I do this faster? Do I need to be doing this at all? What value am I providing with this work?

Laziness and impatience. We focus so we can go home, and we encourage going home because it forces us to constantly focus. This is how great teams scale.

Assessing Your Own Experience

- When was the last time you reviewed your schedule for things that you're doing but that aren't providing a lot of value for you or your team? Look back on the past couple of weeks. Look forward to the next couple of weeks. What did you accomplish, and what do you hope to accomplish?

- If you are still writing code, how does this fit in with the rest of your schedule? Are you doing it after hours? What's driving you to continue to spend this time?

- What was the last task you delegated to a member of one of your teams? Was it simple or complex? How is the person you delegated to handling the new task?

- Who are the rising leaders of your teams? What is your plan for coaching them to take on bigger leadership roles? What tasks are you giving them to prepare them for more responsibility?

- Does the process of writing, releasing, and supporting code seem to function smoothly on your teams? When was the last time there was a noticeable incident with part of this process? What happened, and how did the team respond to it? How often does the process encounter these exceptional conditions?

- When was the last time you pushed your team to cut the scope of a project? When you cut scope do you cut features, technical quality, or both? How do you decide?

- When was the last time you sent email after 8 pm or on the weekend? Did the person you sent that email to respond? Did you need him or her to respond?

7

Managing Managers

The job expectations for managing managers are not that different from the expectations of managing multiple teams. You're responsible for several teams, for overseeing the health of those teams, and for helping them set goals. The difference is one of magnitude. The coverage area for these teams has increased, and there are more projects and people than you could possibly handle by yourself. Instead of managing a couple of closely related teams, you may manage a larger scope of efforts. You may manage functions in your division that you haven't managed before and that you don't have a lot of expertise in—for example, a software engineering manager who now also manages the operations teams for a division.

While managing multiple teams can be exhausting and daunting, managing managers adds a whole new level of complexity that is often a surprise. Consider this email I once sent to my leadership coach:

Managing managers, how do I do it without taking up all my time? What processes should I be putting in place to get appropriate communication out of them and enable myself to scale? How do you help with problems that you aren't in the room to see, with unreliable witnesses? I'm spending all of my time two levels deep in people problems and it's exhausting.

The answers are even less at your fingertips than they were before. Things are now obscured through an additional level of abstraction, and it's easy to miss out on details because you no longer engage regularly with all of the individual developers on each team.

This is a tough growth point. You're going to be pulled in many directions, and figuring out how exactly to spend your time to maximize your leverage across your teams will be critical. In order to do this well, you'll need to practice honing your instincts, and this practice will require you to follow through on things that you're not sure are actually important, but you just sense are off.

Let's take the case of managing the team that is doing work outside of your skill set. It's tempting to just let them roll and only step in if there are problems. However, as a first-timer in this role, you're probably not going to detect problems until they're far gone. You haven't yet built up the discipline or instincts to let yourself intuitively sense where and when to dive in deep, so you need to do so more frequently, even when things seem to be going well.

You'll get a whole new sense of your strengths and weaknesses as you work at this level. People who are good at managing a single team, or even a couple of related teams, fall apart when asked to manage managers, or teams that are outside of their skill set. They're unable to balance the ambiguities inherent in their new role, and fall back into things that they find easy. Sometimes this reveals itself as falling back into spending too much time playing individual contributor. Sometimes it shows up as a person playing project manager instead of training their managers to do that job themselves.

There are people who, by virtue of luck and some skill, get to this level without having to sweat too much. But this is a new game, and it requires a different level of discipline than what's required to manage a team directly. I've discussed getting uncomfortable before, but this is a place where you need to find your discomfort, chase it down, and sit with it unblinking for a while. Here, you need to follow up on all the little things until you figure out what you don't need to follow up on. Is recruiting happening? Are your managers coaching their teams? Has everyone written up their goals for the quarter? Have you reviewed them? What is the status of that project that should be finishing up? That production incident that happened the other day—did the postmortem happen? Did you read the report?

It's so easy to take this position and assume that it's just more of what you were doing before, but that's a mistake. This position is the first level in a much bigger game, the entrée into senior leadership and upper management, and that will require a large number of new skills.

In this chapter, we'll discuss some of the keys to successfully overseeing an entire division, including:

- How to get information from your skip-level reports

- What it means to hold your managers accountable

- Managing first-time and experienced managers

- Hiring new managers
- Figuring out the root of organizational dysfunction
- Cultivating your teams' technical strategy

Ask the CTO: The Fallacy of the Open-Door Policy

I've told my team that I have an open-door policy. They can come to me whenever they want to discuss problems. I even try to hold office hours for them to schedule time! And yet they aren't coming, and I keep finding out about issues that no one has brought to my attention. Why won't my team help me out here?

One thing that managers have to keep in mind is that part of their job is to ferret out problems proactively. There exists an idea that if you make yourself accessible, hold office hours where anyone can meet with you, and tell the team that you are always available, people will naturally bring their problems to your office. There's no need to seek them out, because your team trusts you enough to come to you when things are going wrong.

Except this basically never happens. The open-door policy is nice in theory, but it takes an extremely brave engineer to willingly take the risk of going to her boss (or especially her boss's boss, etc.) to tell him about problems. That even assumes that the engineer knows what the problems actually are well enough to explain them! Even on teams you build yourself, where there's a huge level of trust and respect, some problems will just never get escalated to you. Some of these problems will cause people to leave, projects to get delayed, failures to explode. You turn your back, and the next minute a team that seemed fine has crumbled.

The risk of relying on an open-door policy increases the further away you get from a team. This culminates in the most classic, clueless executive move of relying on office hours instead of meeting one-on-one and with teams directly, and wondering why the wonderful management staff isn't managing to retain great talent or get things done. Some people are great at managing up and hiding problems in their organizations, and you won't ever see these issues if you never take the time to look.

As you manage managers, you'll ultimately evaluate them on the performance of their teams. If the team is not doing well, what then?

Predicting problems is part of your job, so being blindsided by a team that falls apart, has major attrition, or fails to ship a major project on time reflects poorly on you as the higher-level manager. These problems are very expensive to fix the longer they go on, and they won't bring themselves to your doorstep.

So, part of the job is simply to make sure that your 1-1s have room for real conversation beyond a script or a set of to-dos, as I mentioned earlier. But beyond that, you must make the time to proactively hold skip-level meetings with the people who report to your direct reports.

Skip-Level Meetings

Skip-level meetings are one of the critical keys to successful management at levels of remove. And yet many people skip or undervalue them. I know, I've been there. No one wants to add yet more meetings to their calendar, especially the type that are often without an agenda. Still, if you want to build a strong management team, understanding the people who report to those managers and maintaining a relationship with them is hard to avoid.

What is a skip-level meeting? Put briefly, it is a meeting with people who report to people who report to you. There are a few different ways that people hold these meetings, but their purpose is to help you get perspective on the health and focus of your teams. However you choose to hold them, keep this purpose in mind.

One form of skip-level meeting is a short 1-1 meeting, held perhaps once a quarter, between the head of an organization and each person in that organization. This tactic accomplishes a couple of things. It creates at least a surface-level personal relationship between you and everyone in your organization, which keeps you from viewing them as "resources" instead of human beings (something that is a risk in managing large organizations). It also gives those individuals time to ask you questions that they may not feel are worth scheduling a meeting themselves to ask. These meetings are most successful when you provide prompts about potential topics, and remind the person that the meeting is largely for his or her benefit. Each person should come prepared to focus on what he or she is interested in talking to you about.

Some suggested prompts to provide the person you are holding the skip-level 1-1 with include:

- What do you like best/worst about the project you are working on?
- Who on your team has been doing really well recently?
- Do you have any feedback about your manager—what's going well, what isn't?
- What changes do you think we could make to the product?
- Are there any opportunities you think we might be missing?
- How do you think the organization is doing overall? Anything we could be doing better/more/less?
- Are there any areas of the business strategy you don't understand?
- What's keeping you from doing your best work right now?
- How happy (or not) are you working at the company?
- What could we do to make working at the company more fun?

The 1-1 doesn't scale forever. If a quarter has, say, 60 working days, and you have 60 people in your organization, doing one a quarter with each member means you have one every day, or five a week for 12 weeks. This gets worse and worse the more people there are in your organization, and at some point it doesn't make sense—at 1,000 people you'd be doing nothing but these 1-1s, assuming you worked 40 hours a week. However, if you have a smaller organization, dedicating time every quarter to each person does have some benefits.

If you have a larger organization, or are impatient with the idea of adding more unstructured 1-1s to your schedule, there are other ways to get skip-level time. I used to hold skip-level lunches with whole teams, where I would buy lunch for the group and we would talk about whatever was going on. I tried to do these a couple of times a quarter for each team. This has many of the benefits of the 1-1 meeting, in terms of making you more familiar to the team members and vice versa. It doesn't give you the focus to give career coaching to individuals, but it does help you get a sense of the group dynamics and get feedback directly from the teams.

Of course, people act differently in group scenarios, and when you are the Big Boss they may feel uncomfortable complaining in front of others about problems they're having with their manager, even when this person isn't in the room. Many of my lunches served as little more than chats about random technical matters, but I could get a sense for where the team believed their focus needed to be, and I got to answer some detailed questions about the company's strategic focus, the work other areas were doing, or upcoming projects they were interested in hearing more details on.

In the group setting, these questions can be used to draw out information:

- What can I, your manager's manager, provide for you or your team? Anything I should be helping with?

- Is this team working poorly with any other teams, from your perspective?

- Are there any questions about the larger organization that I can answer?

For me, the skip-level lunches provided familiarity, which in turn generated more willingness for people to come to my office hours and cover more sensitive 1-1 topics there, whether by their request or, occasionally, by mine.

The purpose of this skip-level process, beyond maintaining trust and engagement, is to help you detect places in which you're being "managed up" well, to the detriment of the team under that manager. Having people who manage up well in your organization is always a hard situation to detect and respond to. These individuals get to you first, so you hear their perspective before you hear anything else, and you're predestined to think they're in the right and to support their decisions. Skip-level meetings are a chance to hear the other side of the story, to get a reality check from the people on the ground.

At this level, you're constantly making tradeoffs between investing in expensive engagements, such as 1-1s, that can provide deep value but cost you in time and energy, or casual engagements that are more efficient in terms of your time but provide less detailed information. You won't get it perfectly right. There will still be times when you hear too late about a project that's suffering, or a manager who's failing his team, or a team member who's causing problems for others. Invest some time in learning how to maintain these indirect relationships.

Don't underestimate this process, even in the case where you know the skip-level reports well. There's no guarantee that you will keep your close ties with a team just because you used to manage them directly. Managers commonly slip up here when they already have personal relationships and plenty of history working together, so they feel they don't need to do extra work to keep in touch with those teams directly. I've been there and made this mistake. This philosophy sometimes works for a short period of time. But as the teams slowly change, the relationship changes. And even if the team members haven't changed, they won't always come to you with problems they have with their manager. Refer back to "Ask the CTO: The Fallacy of the Open-Door Policy" on page 127 for the reasons why.

Manager Accountability

Whether you have experienced managers or first-timers reporting to you, there is one universal goal for these relationships: *they should make your life easier.* Your managers should allow you to spend more time on the bigger picture, and less time on the details of any one team. This is why they're around. They're more than just people who take some 1-1 meetings off your hands; they are responsible for taking a team of people and helping that team succeed. When they repeatedly fail to do this, they're failing to do their job.

Well, that sounds good, except for one little thing: sometimes managers make your life easier by hiding problems and telling you what you want to hear, until months later you see things falling apart and wonder where you went wrong. So you can't just expect that they'll magically make things better—you have to hold them accountable. This small piece of expertise—learning how to hold managers accountable—will be one of your biggest learning opportunities as you work at this level.

It's hard to hold your managers accountable because accountability on complex teams is often muddled. Your managers may oversee teams with tech leads who are responsible for technical direction and quality. They may also work with product or business managers who set the feature roadmap. And of course it's rare that a team is truly an island, so other teams will have an impact on that organization as well. With all of this responsibility split into different roles, when can you hold a manager accountable?

Here are some tricky but common scenarios I've experienced:

Unstable product roadmap
> The team doesn't feel very productive, the systems are unstable, and there is some attrition happening, but the product organization keeps changing the goals for the team and everything is always an urgent mandate. Is the manager accountable?

Errant tech lead
> The tech lead has been down a rabbit hole trying to redesign one of the core systems. The design doc is still barely started, and work is piling up, but the tech lead insists that this is a big problem that can't be rushed. Is the manager accountable?

Full-time firefighting mode

The manager inherited a team with a bunch of legacy systems that are constantly broken, and the team seems to spend all their time fighting fires. They also support other teams who are using those systems, and the other teams are constantly asking for help and distracting the team with requests. There's a roadmap to move out of these systems, but you haven't heard any reports on the progress against this roadmap, and you know the team is killing themselves to keep things stable and manage the support requests. Is the manager accountable?

The answer to all of these questions is yes. Yes, despite the mitigating circumstances in each case, the manager ultimately needs to take responsibility for pulling the team out of these situations and getting them moving forward, because the manager is accountable for the health and productivity of the team.

When the product organization is constantly changing goals, the manager should identify that the changes are causing problems on the team, and work with product to explain the problem and refocus on what's important. If that fails, she should come to you to help resolve the issue.

When the tech lead is down a rabbit hole, the manager has to bring that person out and work with him to figure out how to make the design process more transparent, bringing in other senior people from other teams if necessary as mentors or collaborators to help him deconstruct the problem and make forward progress.

The manager is responsible for coming to you when the roadmap is stalled because of other issues. If the team can't do anything but fight fires, the manager should put together a plan for tackling the causes of the fires, and if necessary bring requests for hiring more people or adding more people to the team so that they can get the situation under control. When the team is dealing with too much inbound support, the manager is responsible for triaging that support burden and figuring out whether to refuse some of the requests or, again, whether the team needs more people to manage the workload.

In many of these cases, you'll need to help your managers. Sometimes they don't have the clout to push back against product and need your support. They may need your help finding other senior people to partner with their tech lead. You'll probably have to approve any requests for more people to fight fires, or support them in shifting the support burden to other teams. They've done the hard work of identifying the problems that are slowing down their teams, but you need to then help find the solutions or support the path forward. This is what making your job easier looks like—not hiding information, but bringing you clear problems before they turn into raging fires.

Managers need coaching and guidance in the same way that individual contributors need coaching and guidance. Don't forget to spend time with your managers, get to know them as people, and pay attention to their strengths and areas for development. There's plenty to talk about in your 1-1s related to schedule and planning, but make time for feedback and coaching. These individuals will have the biggest impact on your overall organization's success or failure, and in turn will make you look good or bad depending on how well they perform, so take an active role in their management performance.

Good Manager, Bad Manager: The People Pleaser

Marcus is everyone's best friend. He has a team of devoted employees who think he's the best manager in the world. He spends much of his day in 1-1s with everyone from his direct reports to the newest junior hires. Everyone agrees that he makes plenty of time for anyone who wants it, and will listen for as long as you need him to. Bring him any problem, and he promises to make it right. Since he took over the organization, you feel like your concerns are really being heard. However, he never seems to get around to fixing those problems. The colleague you complained about still got promoted. The product team is still railroading you. The goals still don't make any sense. But Marcus is so busy, you can't blame him; after all, there are just so many problems he's dealing with.

Maria is less well loved. She'll make time for you if you request it, but unless you report to her directly, she tends to keep her distance. She can be brusque at times, and doesn't seem to have a lot of patience for office gossip or wasted time. But since she took over the department, things have changed. The roadmap has fewer objectives and they all make sense. Your difficult colleague seems to have gotten some feedback and he's started listening to your ideas. Meetings run better, and the team is focused for the first time in ages. There are still problems, but they seem less important now that you're really getting stuff done. And the most amazing thing is that she seems to go home at a reasonable hour every night!

Marcus is a people pleaser. He has a deep aversion to ever directly making people he cares about unhappy. So if you're in the group that he cares about and you ask him for something, he'll always say yes, even when there's too much happening and he can't possibly deal with all of it. By trying to make everyone happy, people pleasers often burn themselves out.

Signs you might have a people pleaser on your hands:

- Her team loves her as a person but is increasingly frustrated with her as a manager, because she hides problems from them and tries to shield them from the external world.

- He's more interested in having a team that runs smoothly and avoids mistakes than he is in getting the team to push themselves to really become excellent.

- When she is feeling bad, she wears it on her face and it makes the whole team lose confidence.

- He never pushes back on work, but often has many outstanding tasks and excuses as to why none of them has been finished yet.

- She overpromises and underdelivers, and never seems to be able to learn from this experience to promise less in the future.

- He says yes to everyone and sends contradictory messages to both his team and his external partners about what can be accomplished, resulting in widespread confusion.

- She seems to know about all of the problems that are happening in the company, but hasn't directly addressed any of them.

I've seen many versions of people pleasers over the years. One version is the team pleaser, like Marcus. People tend to love him because he spends so much time with them. He wants to engage with your emotions, to make sure that you're happy, and to hear about anything that is bothering you so that he can try to set it right. He doesn't play favorites, but those who are willing to pour their hearts out to him end up getting most of his time. The therapist people pleaser can inspire huge loyalty on his team because of his willingness to hear what you are upset about, and his genuine concern for your emotional well-being. Unfortunately, this can mean he amplifies drama and negativity, and disappoints his team by making promises to them that he cannot possibly keep.

On the other side, there is the external pleaser. She very much wants to make her boss and her external partners happy, and is terrified of revealing problems on her team. As a result, she spends a great deal of energy managing upward and outward, and in particular tends to significantly overcommit her team. Despite wanting to please, she frequently doesn't provide much praise or feedback to her teams internally. This may seem surprising, but this people pleaser struggles to have difficult conversations internally so she avoids talking about serious issues within the team, and this can actually lead to a refusal to acknowledge good work as well as problems. She will never willingly share problems with her manager, and readily agrees to any requests that come in for projects.

In both cases, the people pleaser struggles to say no, and sends contradictory messages to both the team and the external parties. One manager might insist on jumping on every single issue brought in front of the team and doing all of the tedious work of, say, resolving data problems that are due to bugs in the product. Because the manager doesn't actually have the focus for this task, the issues are resolved slowly. Furthermore, the team lacks transparency into the problems their customers are facing, and so they struggle to prioritize fixing issues. By attempting to shield the team from having to do unpleasant work, the manager makes that work drag on and reduces the team's ability to solve problems for good.

People pleasers who focus externally can be a huge blind spot for their managers: because they're so focused on only talking about good things and saying yes to everything that comes their way, their managers often don't even know about problems on the team or within projects until it's too late. These people can be very good at distracting you from your concerns. They have plenty of excuses. They promise to do better next time. They may even be genuinely contrite when you provide corrective feedback, but it's very hard for them to do things that visibly make others unhappy. And you probably like your people pleasers very much as people. They're very nice!

You might think people pleasers create teams that feel safe to be vulnerable and fail, but in fact the opposite is true. These managers make it hard for the team to fail in a healthy way, because of the manager's own fears of failure and possible rejection. An externally focused people pleaser shuts down honest conversation by evasion and, if necessary, emotional manipulation that rests on his status as the person that everyone likes so much. A team pleaser sets her team up to fail by promising things that aren't realistic, and the result is often a team that feels extremely bitter toward either the manager or the company for failure to live up to these inflated expectations.

What should you do if you find yourself managing someone in this situation? Help the person feel safer saying no and externalizing more decisions so he doesn't take failure personally. Providing him with strong partners who can take on the task of determining the work roadmap is a good option. Sometimes people pleaser managers work well in agile frameworks because the team itself takes ownership of work planning. Create better processes for getting work scheduled that don't rely entirely on the manager's discretion. When it comes to promising things to the team itself, having a structure that specifies the requirements for getting promotions or accessing other opportunities can apply here as well. For example, when promotions involve more than the manager's discretion, the people pleaser can rightly point to the process as something outside of her control.

When you're managing a people pleaser, one of the best things you can do is show the person that he's exhibiting the behavior, and highlight the downsides. Sometimes all it takes is awareness that his habit of saying yes is a problem for the team. Recognize that this usually comes from a personal value of being selfless and caring about others, and honor these values even as you try to correct the unhealthy behaviors. People pleasers, after all, just want to make you happy.

Managing New Managers

We talk about management being a career change for engineers, so it's no surprise that first-time managers need a lot of coaching. As you may remember from your first time managing a team, you don't know what you don't know. You probably did whatever good managers had done with you in the past, if you had a good manager to emulate. Perhaps you got a little training or read a book like this one, but more likely you muddled your way through it. Unless, of course, you had a good manager yourself helping you learn the ropes.

Spending quality time with your new managers is important, and you should expect this to be an up-front cost that pays long-term dividends for your organization. You may think that because a new manager has people skills she'll automatically be good at the job. The new manager may believe this as well! But you know there are a number of skills to being a good manager, and even people with solid people skills will need some training to get there.

When you've hired or promoted a new manager, you're often eager to let her loose completely over her team. Finally, that team is no longer your direct concern! Unfortunately, your new manager can be shockingly clueless as to even the basics. Running 1-1s, for example, is an intimidating experience the first time you do it. What do you talk about? How do you give feedback? How do you keep track of takeaways? No book or training can replace you spending some time asking your new manager how her 1-1s are going, and seeing what questions or challenges she may need help with. Sometimes, you just need to remind her to hold them in the first place!

In the face of a new and daunting job, some people just won't do it. When your new manager doesn't manage and slips on too many management details, her team starts to suffer, which means that you start to suffer. When people start quitting because their manager hasn't given them a career path or isn't inspiring them, it's ultimately your responsibility. Use skip-level meetings to help you detect areas where you need to support your new manager fully, and let her know that you'll be holding skip-levels frequently as you help to guide her most effectively.

One common sign of a struggling first-time manager is overwork. A new manager who is working all the time is probably failing to hand off her old responsibilities to other people on her team, and so she's attempting to do two jobs at once. It's one thing for her to be a bit busier, especially as she gets the hang of the new responsibilities, but it's another thing to see her coming in early, staying late, and writing emails all weekend. It's amazing to me how many people never quite learn how to let go of tasks and so are just constantly working longer and longer hours. Make it clear that you expect the new manager to hand off some of her old work, and help her identify opportunities to do so.

Overwork is also often a sign of another new manager danger: the person who thinks she's now in control, the taskmaster of the team, responsible for making all of the decisions. Managers who neglect the job are bad, but managers who take to the job with gusto because they believe it's the key to realizing authority are sometimes even worse. A manager on a power trip domineers her team, and a skip-level meeting with more senior members of the team will reveal their frustration that they have no ability to make decisions themselves. This is slightly different from but related to the micromanager, who expects detailed reports from every member of the team at all times. The micromanager annoys her team to death by asking for an unnecessary level of detail. The control freak takes away the team's ability to make any decisions and views her job as assigning specific work to people to be done. Control freaks usually have bad relationships with their peers in product management and other tech teams, because they often fight to make decisions alone instead of collaborating. What's worse, control freaks often want to hide what they're doing from their manager for fear of having that control taken away. If your new manager is skipping your 1-1s or evading questions about what the team is working on, you may have a control freak on your hands.

The new manager you're training should be ultimately making your job easier by more than just taking the responsibility of doing a bunch of 1-1 meetings off your shoulders. She also needs to be on top of the team's performance and delivery, guiding them to focus on their goals and deliver results. Sometimes new managers fail to realize that they are now responsible for this delivery, and believe themselves helpless in the face of challenging goals or product roadmaps. It's not your job to nag the new manager, to remind her of what she's committed to do, or to hold her hands through the basics of team planning every time it needs to be done, but you will need to coach her through this process at first. Clearly set the expectation up front that you'll hold her accountable for the team, and help her build the skills to achieve this.

First-time managers are tricky because if they truly don't have the willingness to learn and aptitude to become solid managers, they're a big problem. Making the wrong person a manager is a mistake, but keeping her in that position once you've realize she's wrong for it is a critical error. I am hugely in favor of making engineers who wish to go into management take baby steps of mentoring and managing very small teams, but this is not always possible and doesn't always shake out problems that come with scale. Control freak managers, for example, don't often show up as clearly in smaller management situations, instead holding that impulse back until they feel they have the true authority of title. Keep an eye on your new managers. You may need to provide not only coaching but strong corrective feedback in the first six months.

Beyond the coaching that you'll need to provide your first-time managers, I recommend seeking out additional external training. If your HR team has a curriculum for new managers, make sure yours are given the time to attend, and encourage them to do so. You may also seek out additional training opportunities outside of your company, such as conferences that focus on technology leadership, or programs run by current and former engineering managers to specifically address technical leadership topics. New managers are usually eager to learn the tricks of management, and professional programs can help get them up to speed.

Managing Experienced Managers

Now let's move on to experienced managers. This is a very different set of challenges. Experienced managers can be awesome. The right experienced manager knows what needs to be done and does it without needing help from you to get there. He's comfortable with the basics and even has some of his own unique tricks. All good, right?

Of course, there can be major downsides. Management tends to be a very culture-specific task in a company. I can give you best practices all day, but if you either work as a manager or hire a manager for a company that's not a good culture fit, you'll have problems. There is a reason that many young companies want to seed their management teams with people who've been there from the early days and understand the company's DNA. They get the culture, they understand deeply what is important, and they have the internal networks already built to get things done.

So, the first challenge is making sure this person fits in with the culture of your team. We talk a lot about culture fit for all hiring, but managers create subcultures, and a manager who creates an incompatible subculture can be a problem if you want your teams to work together well. Let's say you're hiring a manager because he has expertise in building a certain type of product, and your company lacks expertise in this area. This kind of hire can be great for bringing in knowledge and perspectives. However, often we overvalue expertise in product areas and allow it to blind us to cultural and process fit with our companies and teams. A person with deep expertise in building enterprise-scale warehouse software may seem great on paper to run warehouse technology at your logistics startup. But he won't necessarily work well with an agile, on-site team if he's also used to shipping software once every six months and working only with remote development teams who are not involved in the product ideation process.

If you're building a dynamic, product-centric engineering team, you need managers who understand how to work with teams who ship software frequently, who are comfortable with modern development process best practices, and who can inspire creative product-centric engineers. These skills are so much more important than industry-specific knowledge. It's easier to gain access to industry information than it is to retrain someone who doesn't know how to work in your culture. Don't compromise on culture fit, especially when hiring managers.

Experienced managers will have different ideas about management than you do, and you'll have to work out the differences. Working out these differences, however, is different than letting the manager do whatever he thinks is best. Even (or perhaps especially) if he's been doing it longer than you, be willing to learn from him but don't be afraid to provide your own feedback. Collaborate on areas of difference, allow him to teach you things, and take an active role in the process.

Again, this is a matter of culture. You're responsible for cultivating the culture of your organization, and especially when you've been at the company for a longer period of time, you should ensure that all of your managers respect and nurture the type of culture that you think is best for the team. If you want teams that operate with transparency, make sure the manager shares information. If you want teams that encourage exploration, make sure the manager schedules time and space for his team to explore ideas. Think about what your culture values, and help your managers embody those values while still respecting that every team will be a little bit different and every manager will have certain strengths and weaknesses that you'll need to account for.

How do you inspire experienced managers? The difference between an experienced manager and a new manager is that the experienced person should be capable of managing independently. This means that a lot of the coaching you provide will be less about the nuts and bolts of management and more about how he can have a larger impact on strategy and direction setting for his area. Don't forget to think about tasks that you can delegate to him, and he should be an important advisor when it comes to setting organizational direction. While they may not need as much training as new managers, experienced managers often need help expanding their network both within the company and externally, so look for programs that can help them meet new peers.

Hiring Managers

Your organization is struggling. You've hired in 10 engineers, each with fewer than 3 years of experience. And despite your efforts, none of the existing engineers who might be qualified wants to take on a management role. None of them has much experience managing, anyway, so you would have to do a lot of training to get them up to speed. So, drowning in people, you decide it's time to hire in a new manager to take over some of the team. But how do you do that?

Many people are very reluctant to hire in management from outside, and for good reason. We're barely capable of determining if an engineer is capable of writing good code in a team setting without driving the other team members crazy. And coding is at least a skill that we can ask people to demonstrate for us. Management is...well, what even is it? How do we interview for it? What do we need to watch out for in the management hiring process?

Hiring for managers is a multipart exercise, and those parts are actually very similar to those of a good engineering interview process. First, make sure that the person has the skills you need. Second, make sure that she's a culture match for your organization.

The biggest difference between a management interview and an engineering interview is that managers can, theoretically, bullshit you more easily. The skills of a manager, as we have discussed at length, are pretty much entirely based around communication. Someone who communicates well in a management interview, who talks a good game, can still come in and get nothing done. But engineers who code well in interviews also sometimes fail to ship anything once they join a team. Separate your fear of what happens after you hire the manager from what you're trying to evaluate when interviewing her. You *can* evaluate her and get worthwhile information from the management interview. So how do you do that? Look at the skills you expect from a manager, and ask her about them.

Let's start with 1-1s. As we've discussed, 1-1s are an essential tool for a manager to determine the health of her team and gather and impart valuable information. Any manager you hire should role-play a few 1-1s as part of the interview process. One of the best ways to do this is by asking the people who would report to the new manager to interview her by asking her to help with problems they have right now, or have had in the recent past. Similar to a senior engineer being asked how he might approach debugging an issue that you just resolved, a good manager—even without a full understanding of the people or projects involved—should have good instincts for questions to ask and suggested next steps that might improve matters. You can take it a step further and actually role-play other types of difficult situations, like dealing with an employee who is underperforming, or delivering a negative performance review.

Importantly, a manager must also be able to debug teams. Ask the manager to describe a time when she ran a project that was behind schedule, and what she did in that scenario. Or ask her to role-play with an employee who is thinking about quitting. Ask the manager to describe how she's coached employees who were struggling, and helped great employees grow to new levels.

Ask her about her management philosophy. If she doesn't have one at all, that might be a red flag. While a new manager may not be able to answer this question well, an experienced manager who has no clear philosophy is a cause for concern. What does she think the job of a manager is? How does she stay hands-on, and how does she delegate?

Depending on the seniority, you might have a candidate come in and give a presentation to a group of people. The point of this is not specifically to judge the contents of the presentation, but to see how she is at commanding a room, answering questions posed by a group, structuring her thoughts, and getting up in front of an audience. These are skills that a senior manager should possess, and if she lacks those skills, take that into consideration when you're deciding whether to hire her. I'd caution you not to overvalue this step, however. As a pretty accomplished speaker myself, I think speaking skills are useful for certain types of leadership but not all, and there's only so much you can learn from how well a person presents herself in front of a group. Plenty of otherwise excellent managers are uncomfortable speaking in front of strange audiences.

What about technical skills? You want to make sure that you get enough of a sense of a candidate's technical skills that she'll be able to establish credibility to the team she'll be managing. In the case of someone who will need to write some code, give her an abbreviated version of your standard technical interview. For a noncoding manager, ask technical questions that you believe she should be able to address given her experience. Design and architecture questions based on the types of systems she's built or managed are a good approach. Make sure she can discuss the tradeoffs that were made and why. You might also have her mediate a technical debate between engineers who disagree on the solution to a problem. A good technical manager will know what kinds of questions to ask that tease out the core issues and guide the group to a solid consensus.

So, those are some ideas for skills to look for. The second aspect is cultural fit. As we've covered, this is important throughout your team, but by far the biggest place where it can cause grief is in a management hire. Have you ever worked with a manager who didn't understand the culture of the company or environment? Say, a person from a big company at a startup, who doesn't seem to embrace the startup speed and informality? Or someone from a startup working at a big company, who doesn't know how to get consensus? I'm not suggesting that big company employees can't make great startup managers (look at me, for example), or that startup employees can't succeed in a larger environment, but you want to understand the culture of the company around you and evaluate the manager's ability to fit into that culture.

How do you screen for cultural fit? I discuss this more in Chapter 9, but to summarize, first you need to understand the values of the company around you. Do you have an informal structure that doesn't lean too much on hierarchy, or is hierarchy taken very seriously? Either of those cultures will cause problems for a person who is used to the other. I've seen managers from big companies who treated their peers well but their underlings and other lower-level staff members like they were less than human, which caused massive friction in the startup space. I've also seen managers from startups who were used to being able to act on whatever they thought was most important struggle in environments that required more sign-off from other parties. This is the most obvious cultural element. If you value servant-leadership and you hire a manager who wants to dictate exact marching orders to the team, there will be a bad fit. Similarly, if you value collaboration and hire a manager who thinks that the loudest voice in any conversation should win, you will also have problems.

Culture fit is so important in managers because they shape their teams to their culture, and they hire new people based on their cultural ideas. If you hire a manager who doesn't fit in culturally with the team she's managing, one of two things is likely to happen: the manager will fail and you'll have to fire her, or most of the team will quit and then you may still have to fire her. Sometimes changing the culture of an area is inevitable, and hiring in a new manager will hasten that change. You can use management changes to your advantage in this way. In fact, you see this frequently at growing startups, where they hire in more seasoned managers and executives to round out the lack of experience of the rest of the team. Sometimes this works incredibly well, and sometimes it's a massive failure. No matter what, you will usually see attrition happening around the hire of these bearers of new and different culture, so proceed here with caution.

In his book *High Output Management*,[1] Andy Grove talks about cultural values as one of the ways that people make decisions inside of highly complex, uncertain, or ambiguous circumstances where they value the group interest above their own. I find this insight very powerful. His observation is that most new hires act in self-interest until they get to know their colleagues, and then they move into group interest. So, if you start them in a highly complex or uncertain job, they tend to fail unless they quickly settle into the cultural norms and use cultural values to align their decision making. If you can screen for managers who naturally gravitate toward the cultural values that your company already possesses, they are more likely to make this shift quickly than managers who have very different personal beliefs.

Finally, I would be remiss if I did not point out one of the critical elements to hiring in new managers: the reference check. Do thorough reference checks for anyone you're planning to bring on board, even if you've worked with that person before. Ask the references to describe the ways that the person succeeds as well as the ways she fails. Ask them if they would work with or for this person again.

---

1 Andrew S. Grove, *High Output Management* (New York: Vintage Books, 1983).

Ask them what they love about the person, and what drives them crazy. If you're not doing reference checks when you hire management, you're doing your team a massive disservice. Reference checks, even ones chosen carefully by the candidate, often reveal a lot about what you can expect to get when hiring her. Don't leave out this crucial step.

Ask the CTO: Managing Outside Your Skill Set

I'm now responsible for managing not only my division's software development teams, but also the operations and QA teams. I have never had to manage these types of teams before, so do you have any advice on doing it well?

Be careful! It's easy to think of this as a small leap from managing other software developers, but in my experience, you need to track different important details than you're used to in these areas, and if you've never managed these types of teams before, it's hard to know which details to focus on. Unfortunately, it's very easy to miss out on problems in unfamiliar areas until it's too late.

What happens when this situation goes poorly? In my personal experience, it can be a big problem. When you hire a manager over a team doing things that you don't deeply understand, it's easy for that manager to go the wrong way for a very long time before you realize what's happening. This is particularly difficult when it comes to any sort of projects that have long timelines, where it's easy to hide a lack of progress.

An interesting way to combat this problem is to use the same mindset that I advised when we talked about mentorship—namely, being very curious. Remember that you're not expected to know everything just because you're a manager. Use this to your advantage. Ask the person to teach you about the work she does. Sit down with her and treat her as if she were your mentor, the person to teach you the ropes for this job. Whether it's QA, design, product management, or technical operations, ask lots of questions, but in an open way. Make it clear to the person that your goal is to understand what she does so that you're capable of appreciating it better.

One other piece of advice here is that, while you may be tempted to spend more of your time in the areas that you are most comfortable with, to be prepared to devote significant time to the areas new to you, especially in the beginning. It's tempting for managers who want to trust and delegate to simply assume that people will do the right thing and let them go, but this can easily result in you missing problems in these areas for far too long. What's worse, if you view these areas as uninteresting or unworthy of your time, you may find yourself reluctant to deal with problems in these teams even when people are clearly drawing attention to them. You'll feel guilty for having ignored the areas in the first place, and your natural aversion may keep you from facing up to issues for far longer than you otherwise would allow. Grit your teeth and make the time to get comfortable with each area; take time to get to know the manager and employees in the team, and practice asking for details about the area, so that you can start to learn and develop a sense for what the people in that team are actually doing.

Debugging Dysfunctional Organizations

I believe that the best engineering managers are often great debuggers. Why would this be? What is it about these two tasks that produces such an overlapping skill set?

A great debugger is relentless in his pursuit of the "why" for a bug. This is simple when we're looking for errors in application logic, but we all know that bugs can go many layers deep, particularly in complex systems that involve many separate parts operating over time-delayed networks. A sign of a poor debugger is a person who, when he adds a log statement to a piece of concurrent code to attempt to find an error and sees that the error can't be reproduced, assumes that he's fixed the problem. It's a lazy habit, but a common one. Sometimes there are problems that seem impossible to determine, and many people don't have the patience to dig through layers of code (theirs and others'), logfiles, system settings, and whatever else is needed to get to the bottom of something that happened only once. I can't blame them. Obsessive debugging of one-off issues is not always a great use of your time, but it does show a mind that won't be satisfied with the unknown, especially when that unknown might cause you to be paged at two o'clock in the morning.

What does this have to do with management? Managing teams is a series of complex black boxes interacting with other complex black boxes. These black boxes have inputs and outputs that can be observed, but when the outputs aren't as expected, figuring out why requires trying to open them up and see what's going on inside. And, just as sometimes you don't have the source code, or the source code is in a language you don't understand, or the logfiles aren't readable, the black boxes of teams can resist yielding their inner workings.

Let's work through an example. You have a team that feels slow. You've heard complaints from their business partners and product manager that they're slow, and you agree that the team just seems to lack the same energy as your other teams. How do you figure this out?

HAVE A HYPOTHESIS

To properly debug a system, you need a reasonable hypothesis that explains how the system got into the failed state, preferably one that you can reproduce, so that you can fix the bug. To debug a team, you also want to look for a hypothesis around why the team is having problems. Do this in as minimally invasive a way as possible, to prevent your meddling from obscuring the problems. As an added challenge, team problems are not generally single failures but are more like performance issues. The system is running, but it seems to slow down from time to time; the machines are OK, except occasionally they crash; people seem happy, but attrition is too high.

CHECK THE DATA

Debugging a team deserves the same rigor you would apply to debugging a serious systems issue. When I debug a systems issue, the first thing I look at is logfiles and any other record of system state from the time of the incident. When you're looking at a team that isn't producing work fast enough, look at the records. Look at the team chats and emails, look at the tickets, look at the repository code reviews and check-ins. What do you see? Are production incidents happening that are taking up lots of time? Are a bunch of people sick? Are they bickering over coding style in their code review comments? Are the tickets that are being written vague, too big, too small? Does the team seem upbeat in their communication style, sharing fun things as well as important work in chat, or are they purely business? Look at their calendars. Is the team spending many hours a week in meetings? Is their manager not doing 1-1s? None of these things are necessarily smoking guns, but they may point to an area to address.

OBSERVE THE TEAM

Perhaps everything seems OK in all of these indicators, but the team still just is not performing as well as you believe they should. You know the talent is there, the team is happy, and they're not being overburdened by production support. So what's happening? Now is the time to start doing some potentially destructive investigations. Sit in their meetings. Are they boring to you? Is the team bored? Who is speaking most of the time? Are there regular meetings with the whole team where the vast majority of the time is spent listening to the manager or product lead talk?

Boring meetings are a sign. They may be a sign of inefficient planning on the part of the organizers. There may be too many meetings happening for the information covered. They may indicate that the team members don't feel they can actually help set the direction of the team, or choose the work that will happen. They often signal a lack of healthy conflict on a team. Good meetings have a heavy discussion element, where opinions and ideas are drawn out of the team. If the meetings are overscripted, so that no real conversation can take place, it stifles that creative discussion. If people are afraid to disagree or bring up issues for fear of dealing with conflict, or if managers always shut down conflict without letting disagreements air, this is a sign of an unhealthy team culture.

Be aware, though, that while teams can be black boxes, they share a characteristic of another famous box—the one containing Schrödinger's cat. The point of Schrödinger's experiment was to show that the act of observing changes the outcome, or rather, causes an outcome to happen. Likewise, you can't go into a team and not change the behavior of that team by being around them, sitting in their meetings, and watching their standups. Your presence changes the team's behavior and may hide the problem you're trying to find, in the same way that a log statement can cause a concurrency issue to be magically erased, at least for some time.

ASK QUESTIONS

Ask the team what their goals are. Can they tell you? Do they understand why those are the goals? If they don't understand the goals of their work, their leaders (manager, tech lead, product manager) aren't doing a good job engaging the team in the purpose of the work. In almost every model of motivation, people need to feel an understanding and connection with the purpose of their work. Who are they building these systems for; what is the potential impact on the customer, the business, the team? Did they have any part in deciding these goals, and the projects that they're doing to achieve them? If not, why not? When you see a team spending all of their time on engineering-sponsored projects and neglecting product/business projects, it's likely that the team doesn't appreciate or understand the value of the product/business projects they're supposed to be working on, and therefore they lack the motivation to tackle them.

CHECK THE TEAM DYNAMICS

Finally, you might take a look at the actual team dynamics. Do people like each other? Are they friendly? Do they collaborate on projects, or is every person working on something independently? Is there banter in the chat room, in emails? Do they have a good working relationship with other adjacent departments, and with their product managers? These are little things, but even very professional groups tend to have a degree of personal connection between the members. A bunch of people who never talk to each other and are always working on independent projects are not really working as a team. There would be nothing wrong with that if the team were performing well, but given that they're not, this may be contributing to your problem.

JUMP IN TO HELP

Sometimes managers of managers choose to approach such problems as something that the team manager just needs to fix. You measure the manager on the output of his team, after all, and it is his responsibility to fix it if things are not going well. This is true, but just as I sometimes jump in and help debug complex system outages even though I rarely write code, it's OK to jump in and help debug team issues as you see them, particularly when the manager in question is struggling. It can be an opportunity to teach the manager and help him grow. It can also reveal more foundational problems with the organization, such as a lack of senior business leadership that even the best managers can't identify or resolve alone.

BE CURIOUS

The pursuit of *why* when it comes to organizational problems is the thing that gives you patterns to match on, and lessons to lead with. We get better at debugging by doing it often, and learn which areas tend to break first and which indicators provide the most value for understanding issues. We become better leaders by pushing ourselves and our management teams to really get to the bottom of organizational issues, searching for why so that we can more quickly resolve such issues in the future. Without the drive to understand why, we rely on charm and luck to see us through our management careers and to make our hiring and firing decisions. As a result, we have a huge blindspot when it comes to truly learning from our mistakes.

Setting Expectations and Delivering on Schedule

One of the most frustrating questions that engineering managers get asked regularly is why something is taking so long. We've all been asked this question before. We've been asked it as hands-on engineers, as tech leads, and as managers of small teams, but the question takes on a whole new level of intensity when you're managing team managers, because answering it is significantly harder when you aren't embedded deeply in the details.

First things first: hopefully you're being asked this question because something is running over plan by a significant margin. That is the time when it makes the most sense to ask, and when you should do your best to understand the scenario and answer.

Sadly, we are often asked this question in times when things are not taking any longer than the estimate. We are often asked this question when our leadership, for whatever reason, either didn't like the original estimate or didn't ask for it at all, and now they're upset, despite nothing going wrong.

Therefore, you must always be aggressive about sharing estimates and updates to estimates, even when people don't ask, especially if you believe that the project is critical or likely to take longer than a few weeks. This means you must be aggressive about getting estimates, and as we all know, software estimation is a very difficult process. Negotiating the process that your team uses to estimate, on what timescale, for what projects, may be part of your job at this level.

Engineers often don't want to estimate at all, or estimate beyond the boundary of an agile sprint (generally two weeks). This philosophy is completely reasonable if you believe that estimates must be fairly accurate, that the requirements are unknown or will change frequently, and that most of the work should be bound to features that mostly fit within one or two sprint efforts. With that said, few of these things are always true. Estimates are often useful even if they aren't perfectly accurate because they help escalate complexity to the rest of the team. Not every project necessarily has requirements that change frequently, and it's possible to do up-front work to drastically reduce the unknowns that make software estimation difficult. You may argue that the up-front work sometimes makes the overall process take longer than simply looking at the project sprint by sprint, and you may be right, but again, we're not just talking about engineering teams here. We're talking about businesses that want to plan and get ideas of costs for effort. We're also, in some sense, talking about goal setting and learning how to get better at understanding the complexity of our software and systems. We can't predict the future perfectly, but teaching our teams how to hone their instincts about complexity and opportunity is a worthy goal.

So, accept the fact that you'll need to do some degree of estimation. Play with different methods, and see what works for your company, but make it a habit across your teams.

Another core element of agile software development is the emphasis on learning from the past. When estimates are wrong, what are we learning about unknown complexity? What are we learning about what is worth estimating, when? What are we learning about how we communicated those estimates and who was disappointed by the miss?

It is your job to make it clear, as best you can, what "long" actually is by providing your best view into the timescale of a project, and proactively updating that view when it changes, especially if it gets much slower.

Even with your best efforts, sometimes you'll be asked this question when you've been clear about what "long" is, when you're not actually taking so long, or when things completely outside of your control have come up and caused delays that were well communicated. It sucks when this happens, and it usually happens because someone is feeling stressed or being pushed to deliver faster than you ever claimed to be able to deliver. There's no easy answer to this situation. Sometimes patiently reminding the other person that things are going as fast as they can and everything is on schedule is the only solution. But blame is not usually a fully rational exercise, or one that can be totally avoided under stress. Showing some empathy for the person providing pressure and being willing to help out in other ways can go a long way to shifting focus from blame to action.

Finally, don't be afraid to work with your managers, tech leads, and the business to cut scope toward the ends of projects in order to make important deadlines. As the senior manager, you may need to play tiebreaker and make decisions about which features are worth cutting, and which features are essential to the project's success. Help the team look out for these features, and be willing to take the fall for cutting back on someone's favorite idea if it's essential to getting the larger project done. Be smart about what you're willing to give on. If you give only on matters of technical quality, you'll just slow down the team after the project is launched, so be sure to look at product features in addition to technical nice-to-haves.

Challenging Situations: Roadmap Uncertainty

A very common problem that managers at all levels face is the challenge of changing product and business roadmaps. Especially in smaller companies, it's hard to get people to commit a year in advance to the work that will be done for the next year. Even at big companies, changes in the market can lead to sudden-seeming shifts in strategy that cause projects to be abandoned and planned work to be cancelled.

This is really hard for engineering managers to deal with. Changes in strategy are where being stuck in "middle management" feels the most unpleasant. You may have very little ability to push back on the changes to strategy coming from above, and even when you've promised your team that certain projects will happen, you sometimes have to pull back on that promise because of unexpected changes. This makes the team unhappy, and they complain to you. Because you have no ability to do much about it, you can feel like it exposes you as powerless, and your team might feel that they're being treated not as humans, but as cogs in the corporate machine.

Coming into play here is a secondary challenge: how do you make the time for your team to deal with technical debt and other engineering-focused projects when there doesn't seem to be a clear process for prioritizing that work? After all, if you don't put any time into dealing with the technical issues themselves, your team's ability to do product features will slow down. And yet the product team will never have technical debt on their roadmap, so the planning process often means there is no time allotted for this type of work.

STRATEGIES FOR HANDLING ROADMAP UNCERTAINTY

There are few strategies I've learned about building a roadmap:

- **Be realistic about the likelihood of changing plans given the size and stage of the company you work for.** If your startup has a history of changing the year's plans every summer to account for the business results from the first half of the year, be prepared for a change in the summer, and try not to promise things to your team that would require continuity beyond that point.

- **Think about how to break down big projects into a series of smaller deliverables so that you can achieve some of the results, even if you don't necessarily complete the grand vision.** Breaking down the technical work will require you to work closely with the product or business managers to figure out how the details should be prioritized. All of you should be aware by now that things will change quickly, so everything must be repeatedly reexamined with an eye toward what's most valuable right now.

- **Don't overpromise a future of technical projects.** Don't promise your team exciting technical projects "later," because the product roadmap for later hasn't been written yet. This kind of thinking will get hopes up and then disappoint. If the project is important, get it scheduled now—or as close to now as possible. If the project is not urgently important, you can put it on the backlog, but you should be realistic that once "later" rolls around, there will be a long list of competing priorities from other parts of the business. If you haven't taken the time to articulate the value of this work, it will get pushed aside in favor of projects that are more clearly valuable.

- **Dedicate 20% of your team's schedule to "sustaining engineering."** This means allowing time for refactoring, fixing outstanding bugs, improving engineering processes, doing minor cleanup, and providing ongoing support. Take this into account in every planning session. Unfortunately, 20% is not enough to do big projects, so additional planning will be needed to get major technical rewrites or other big technical improvements. But without that 20% time, there will be negative consequences with missed delivery goals and unplanned and unpleasant cleanup work.

- **Understand how important various engineering projects really are.** Product and business projects usually have some kind of value proposition to justify them. However, the same rigor isn't always applied when it comes to technical projects. When an engineer comes to you with an engineering project that she wants to do, think about framing the project by answering these questions:

  — How big is that project?

  — How important is it?

  — Can you articulate the value of that project to anyone who asks?

  — What would successful completion of the project mean for the team?

  The value of these questions is that you start to treat big technical projects the same way as product initiatives. These projects have advocates and goals, they have schedules, and they are managed like other big initiatives. This is a scary process because there are times when you "know" something is important, but you don't know how to articulate it in a way that the business will value. Especially given the complex nature of technical projects and the challenge of measuring things like engineering efficiency, you're sometimes stuck trying to explain technical details to a nontechnical partner who may not totally understand where you're going or why. My advice is to do your best to gather data to support yourself, and talk about what will be possible when the work is done. If you look at a technical project and realize that you're proposing a bunch of work for a system that is rarely changed and won't enable core improvements to your technology or business, it probably isn't worth the effort. Unfortunately, there is never enough time for all the exploratory engineering, legacy code cleanup, and technical quality improvements your team will want to do, and this process will help you pick your battles.

So, back to our uncertain roadmap. Projects change. Teams may even be disbanded or moved around in ways that you don't understand or agree with. As a manager, the best thing you can do is help people feel capable of tying up loose ends, stabilizing the current in-flight projects, and easing into their new work in a controlled fashion. This is an area where you can and should push back. Make sure that your teams get adequate time to finish up current work. Furthermore, push for engineering involvement in the early planning for the new work so that people can get excited about the projects they are moving on to. Take the time yourself to understand the reasons for the move, and even if you don't totally agree, do your part to help make those reasons clear to your team and help them understand the new goals. The calmer you can be in the face of these changes, and the better you can show (or fake) enthusiasm for the new direction, the easier the transition will be for your whole team.

When you are faced with waves, you can let them pull you under or you can learn how to surf. Hang 10.

Staying Technically Relevant

One question I hear a lot from managers is "How do I remain technically relevant?" We know that without investment into our technical skills, we run the risk of becoming out of touch with the field and obsolete before our time. But what does technical relevance really do for you? To answer that question, let's start by clarifying your technical responsibility.

OVERSEE TECHNICAL INVESTMENT

To move forward, systems need constant technical work: new languages, frameworks, infrastructure, and features. There's only a limited amount of development time and energy that can go into improving these systems, and you're accountable for making sure the team is placing its technical bets in the right places. You oversee these investments by matching the proposed technical projects and improvements to the future of the product or customer needs. Looking holistically across the portfolio of projects, you can see where the areas of greatest need or opportunity lie, and focus the team's efforts accordingly.

ASK INFORMED QUESTIONS

You're not the person who identifies all technical projects. Having accountability for the team's technical investments doesn't mean that you personally do the research to find potential investments. Instead, you guide these investments by asking questions. What are the current projects, and what surprises or bottlenecks have they uncovered? How is the team thinking about the future of the systems? Which teams are asking for more engineers, and why do they think they need more people? Which teams are slow but don't want to add more people to improve their throughput? Why are they advocating for this specific project right now? You need to know enough about the work to sniff out misguided efforts and evaluate proposed investments.

ANALYZE AND EXPLAIN ENGINEERING AND BUSINESS TRADEOFFS

By knowing what your teams are excited about and what they value, you can rally them around product initiatives. You know enough to raise concerns when a feature idea is technically difficult and when a technical idea has unforeseen business implications. You ensure that engineers make decisions with an understanding of the business perspective and the future of the product roadmap. When technical work requires uncertain research and development, you're capable of explaining why that uncertainty exists to your nontechnical counterparts. Understanding the business and customer goals, you offer guidance as to which technical projects can achieve those goals within reasonable time frames.

MAKE SPECIFIC REQUESTS

As a director-level manager, you still need to have enough understanding of the technology in your organization to make specific requests without distracting the senior engineers with questions. By knowing enough about the progress of your teams, the projects, and bottlenecks, you can filter out technically infeasible ideas and map new initiatives onto ongoing projects. These specific requests should be used to keep the teams productive and balance technical risks with organizational goals. Here's an example of how this works:

> *Your VP tells you that she wants to improve the search experience to grow active users by next quarter, and she can give you more engineers to do the work faster. You know the team can't usefully add engineers to modify search because it's in the process of being rewritten. Instead, you direct them to prioritize the work to expose the new API earlier so that the product team can finally run some of the tests they've been asking for. You explain what's possible to the VP and make sure the team is focused on finishing work that can make those higher-level goals achievable.*

Managers who don't stay technical enough sometimes find themselves in the bad habit of acting as a go-between for senior management and their teams. Instead of filtering requests, they relay them to the team and then relay the team's response back up to management. This is not a value-add role.

USE YOUR EXPERIENCE AS A GUT CHECK

This is a highly technical job that can't be done by a person who does not understand and appreciate the challenges and tradeoffs of software engineering and technology. If your team invests their time poorly, it will reflect on you as their leader for not helping them come to better decisions. Rely on your instincts to guide where you spend your time and attention, and don't neglect your technical instincts just because you are busy with people and organizational challenges.

Given your level of technical responsibility, how should you invest your time in order to stay technically relevant?

- **Read the code.** Occasionally taking the time to read some of the code in your systems can help remind you what it looks like. Sometimes, it also shows you places where things have gotten ugly and need attention. Looking over code reviews and pull requests can give you insight into changes that are happening.

- **Pick an unknown area, and ask an engineer to explain it to you.** Spend a couple of hours with one of the engineers who is working on something you don't understand, and ask him to teach you about that area. Go to a whiteboard or share a screen and have him pair with you on a small change.

- **Attend postmortems.** When outages happen, make it a priority to attend the post-outage debriefs. These meetings are often full of details about the process of writing and deploying software that you miss when you aren't coding every day. Standards that you thought were obvious have been neglected or ignored. Communication between teams is lacking, and tooling is hurting more than it is helping. In times of failure you can most clearly see where problems have built up, and you learn where your attention is needed.

- **Keep up with industry trends in software development processes.** One major weak spot for managers is losing touch with the tools and processes for actually developing, testing, deploying, and monitoring code. These are the places where new ideas can make your teams significantly more effective. Not every trend is worth pursuing, but make time to learn about how other teams deliver software so you can keep your teams evolving.

- **Foster a network of technical people outside of your company.** The best stories are the ones that come from people you trust. Keeping up a network of peers in engineering and engineering management gives you people to ask for opinions on new trends. Use this network to get the real experiences behind the blog posts, talks, and sales pitches for new technology.

- **Never stop learning.** Find articles and blog posts about technology and read them. Watch talks. Pick something you're really curious about and dig in a little bit deeper, even if it isn't relevant to your team or company. Don't be afraid to ask questions of your team and look for opportunities to learn from them. Learning is a skill that you can practice to keep your mind sharp.

Assessing Your Own Experience

- How often do you talk to your skip-level reports? Do you meet with them one on one, or as groups? How do you proactively reach out to your teams? How much time do you spend seeking out information, instead of passively handling the information that comes to you? When was the last time you sat in on a team meeting?

- Without looking at your existing documentation, write down your view of the job description for the engineering managers who report to you.

  — What are they responsible for?

  — How do you evaluate them?

  — What areas are most important for success, in your opinion?

- Now, look at the job description your company uses. Are there differences in what you wrote compared to that description, or do they match well? Given that description, what things are you potentially overlooking in evaluating them?

- Finally, do a quick mental review of their current performance. What areas need coaching and development? Make time to cover this in your next 1-1.

- If you manage an area that is outside of your technical comfort zone, how often do you check in on that area to make sure things are going well? Have you taken some time to learn from the manager of that area a little bit about what it takes to succeed in that role? What new things have you learned in the past three months that help you understand that team better?

- If you have one team that is clearly operating more smoothly than others, what are the differences you notice in their processes? Their interactions? Is their manager doing things differently than other managers? How does the team interact with that manager, and how does that manager interact with you?

- What is your interview process for managers? Do you spend time talking about their personal values and their management philosophy? Do you have the team interview their potential manager, or do you keep them out of the process? Do you spend time getting references for candidates?

- What are your organization's goals this quarter? This year? How are you merging product goals (if any) with the technical goals? Does your organization have a mandate that is well understood by the teams?

8

The Big Leagues

The day-to-day job of a senior manager depends greatly on the company you're in. It would be silly of me to say that my job running a startup engineering organization of 70 people was the same as the job held by a senior manager who's over thousands of people at a Fortune 500 company. Books upon books upon books are written about senior management at scaled companies from a general perspective. I've listed some recommended reading at the end of this chapter for general-purpose senior leadership advice. All of these books are fantastic and essential guidelines for senior leaders.

But we're not general-purpose senior leaders. We're technology senior leadership. This book is for the engineer who wrote code for a while and eventually moved into management, and successfully grew a career up along that path. And as engineers, we share some common responsibilities that are specific to our role as technologists and come partially from our upbringing working in that ever-changing world.

As technical senior managers, we bring special skills to an organization. In particular, we bring a willingness to embrace and drive change as needed. We're able to question the way we do things now, and try different things if our current way of operating isn't working. We understand that technology evolves quickly, and we want our organizations to evolve to keep up with these changes. We have a unique role, but we still need to succeed in our general senior management roles. It's not enough to be a change agent; we have to create an organization that can successfully follow through on the changes we want to push.

Your first job is to be a leader. The company looks to you for guidance on what to do, where to go, how to act, how to think, and what to value. You help set the tone for interactions. People join the company because they believe in you, in the people you hired, and in the mission you helped to craft.

You're capable of making hard decisions without perfect information and willing to face the consequences of those decisions.

You're capable of understanding the current landscape of your business, as well as seeing into its many potential futures. You know how to plan for the months and years ahead so that your organization is best suited to handle those potential futures and capture opportunities as they come along.

You understand organizational structure and how it impacts the work of teams. You know the value of putting in place management that strengthens this structure rather than undermining it.

You can play politics in a productive way, in order to move the organization and the business forward. You work well with your peers outside of engineering and seek out their perspectives in addressing issues with a wide scope.

You understand how to disagree with a decision and commit to deliver on it even though you disagree.

You know how to hold individuals and organizations accountable for their output.

In his book *High Output Management*,[1] Andy Grove breaks down management tasks into four general categories:

Information gathering or information sharing

Sitting in meetings, reading and writing emails, talking to people one on one, gathering perspectives. The strong senior leader is capable of synthesizing large quantities of information quickly, identifying critical elements of that information, and sharing the information with the appropriate third parties in a way they will be able to understand.

Nudging

Reminding people of their commitments by asking questions instead of giving orders. It's hard for a leader of a large team to forcefully guide that team in any direction, so instead rely on nudging members of the team to keep the overall organization on track.

---

1 Andrew S. Grove, *High Output Management* (New York: Vintage Books, 1983).

Decision making

Taking conflicting perspectives and incomplete information and setting a direction, knowing that the consequences of a poor decision will impact both you and possibly the whole team. If making decisions were easy, there would be much less need for managers and leaders. However, as anyone who has spent a lot of time managing can tell you, making decisions is one of the most draining and stressful parts of the job.

Role modeling

Showing people what the values of the company are. Showing up for your commitments. Setting the best example for the team even when you don't feel like it.

Whether you're a CTO, a VP, a general manager, or a head of engineering, your days are shaped by those four tasks.

What Is My Job?

I came to engineering from what the industry likes to call a "nontraditional background." I suffered (suffer?) a strong sense of impostor syndrome when interacting with folks from more traditional CS backgrounds. This was especially difficult when leading folks whom I perceived to be more technical than I.

This background, combined with a strong desire to be seen as smart and "right," sometimes led to some less-than-productive conversations about technology directions. I'd find myself arguing the merits of a language choice or technology purely based on the technical merits. When this happened I became another engineer arguing in a group of engineers.

It took me a long time to realize that my job wasn't to be the smartest person in the room. It wasn't to be "right." Rather, my role was to help the team make the best possible decisions and help them implement them in a sustainable and efficient way.

I care about technology—it's a factor in every decision we make as a team—but technology alone doesn't make a productive and happy team. A good leader shapes technology discussions to inject the strategic objectives and take into consideration the nontechnical implications of a technical decision. It's not about being the lead engineer, chasing the latest language or framework, or having the shiniest technology. It's about building a team with the tools and attributes to build the best possible product for our customers.

—James Turnbull

Models for Thinking About Tech Senior Leadership

I have a very opinionated view of the job of the CTO, especially as it applies to product-focused startups. I stand by it, but I realize that it's not the universal model for CTOs at all companies. I also know that there's a great deal of muddiness in senior levels of management. Where does the VP of Engineering fit in? How about the Chief Information Officer? Is that a role we need to fill? What about the product team?

Instead of trying to cover all of the variants of roles in senior leadership, I'm going to start by explaining some of the most common roles that senior management can play, and how those roles can fit together. These descriptions may make sense for your company, and they may not. Some people will be able to play several parts, some people will be able to play only one or two, and some companies won't even need people to play all of these parts. All of them break down at large-enough companies, at which point you often have to look at these roles division by division. But by presenting a taxonomy, I hope to help you consider the possible skills you'll need to be successful in various senior leadership positions. The common roles include:

Research and development (R&D)
> Some companies focus on expanding the cutting edge of technology, and therefore may have a senior leader in the technology organization who is focused on experimentation, research, and new technology generation. This role might own technology strategy, or it might be purely a role for finding new ideas.

Technology strategy/visionary
> Technology strategy meets product development. This person often also manages the product organization. He's focused on how technology can be used to grow the business and works to predict the evolution of technology as it applies to the company's industry. This differs from R&D in that the

visionary isn't usually focused on research potential; he uses business and technology trends to guide his decisions.

Organization

The organizational manager guides the structure and people in the organization. She owns the plans for staffing the team as well as organizational structure, ensuring projects are staffed accordingly. This role is often paired with "Execution," discussed next.

Execution

Usually paired with "Organization," this person makes sure things actually get done. He helps align roadmaps, plan work, and coordinate large efforts. He makes sure that projects get prioritized. He breaks roadblocks, resolves conflicts, and makes decisions in order to get the team moving forward.

Face of technology, external

When a company sells software-based products to other companies, one of the senior technology leaders usually is expected to participate in this sales cycle. She may attend client meetings and speak at conferences to encourage use of the product. A company that is invested in building an engineering brand for hiring purposes may also need one of the senior leaders to play this role by speaking at conferences and recruiting events.

Infrastructure and technical operations manager

This is the person responsible for all of the technology infrastructure and infrastructure operations. This role might be cost-focused, security-focused, or scaling-focused depending on the company and stage of evolution.

Business executive

This is someone whose first focus is the business itself. This person understands the industry and understands the other essential functions of the business at a high level. He balances internal development needs with business growth needs and owns prioritization of projects at a high level.

Here are some combinations of these roles that I've observed personally and from a distance:

- Business executive, technology strategy, organization, and execution: CTO or Head of Engineering (VP/SVP)

- R&D, technology strategy, external face of technology: CTO, Chief Scientist, Chief Architect, sometimes Chief Product Officer, usually for a company that is selling a software-based product
- Organization, execution, business executive: VP of Engineering, General Manager
- Infrastructure manager, organization, and execution: CTO/CIO, possibly VP of Technical Operations
- Technology strategy, business executive, and execution: Head of Product (or Chief Product Officer), sometimes CTO
- R&D, business executive: CTO or Chief Scientist, cofounder
- Organization and execution: VP of Engineering, sometimes Chief of Staff

As you can see, organizations can mix and match and define these roles in different ways depending on business needs. The CTO role in particular changes its focus wildly depending on the company, although most strong CTOs have a strategic function under their command, whether it's business-focused, technology strategy, or both.

What's a VP of Engineering?

In an organization where the CTO is the executive manager of all of engineering, responsible for strategic leadership and oversight, what does the VP of Engineering do? What does a great VP of Engineering look like?

The VP of Engineering role varies just as much as the CTO role does, based on the needs of the organization. However, the VP role has one obvious difference from the CTO role. The VP is usually at the top of the management career ladder for engineers. This generally means that the VP is expected to be an experienced manager of people, projects, teams, and departments.

As companies grow, it's common for the VP-level role to change from an organizationally focused job to a business strategy position. These people often act as mini-CTOs for divisions, balancing strategy and management. You may end up with multiple people in "VP of Engineering" roles who are responsible for portions of the engineering team. Their roles become more strategy-focused over time, while the organizational management duties fall more to their deputies at the director or senior director level. Let's put the complexities of large companies aside and focus on the VP of Engineering that you'll most likely encounter in a company that has a single person in this role.

As the person in charge of the day-to-day operations of the team, a good VP of Engineering has a solid handle on processes and details. She's capable of tracking several in-flight initiatives at once and making sure they're all going well. Often a great VP of Engineering is described as having a good "ground game." This person is capable of dropping into the details and making things happen at a low level. While some CTOs will do this, if there's both a CTO and a VP of Engineering, the VP is usually the one pushing the execution of ideas, while the CTO focuses on larger strategy and the position of technology within the company.

The VP of Engineering also handles a significant amount of management responsibility. She aligns the development roadmap to the hiring plan, planning out how teams need to evolve in order to make sense given the expected growth in the team. She may work closely with recruiting and run the hiring process, ensuring that résumé screening and interviewing are going smoothly. She should be a coach to the engineering management team, identifying and improving existing talent and working with HR to provide training and development resources for those leaders.

The VP of Engineering job is both a big one and a detail-oriented one. This is one of the reasons it's such a hard job to hire for, even though most companies will need to hire this role in from outside. This person has to be good at quickly understanding what's going on in an organization. She must be able to gain people's trust and show wisdom in management and leadership. Unfortunately, most engineers are reluctant to trust people without technical credibility, but many managers at this level of seniority won't be interested in going through a hard technical interview process only to take on a role that's mostly focused on organizational management.

The VP of Engineering also needs to have some stake in organizational strategy, and often she owns this strategy entirely. She'll be heavily involved in helping teams set goals to achieve business deliverables, and that means she'll need to be closely aligned with the product team. She must ensure that the roadmap is realistic and that the business goals are translated into achievable goals for the technology organization. She should have a strong business and product instinct and a track record of getting teams to deliver big projects, including the ability to negotiate deliverables.

The people I know who excel in this role are capable engineers who care deeply about their teams and prefer to stay out of the spotlight in favor of creating high-performing organizations. They're interested in the complexities of getting people to work together effectively. They want their teams to be happy, but they know that it's important to tie that happiness to a sense of accomplishment. They represent the health of the team to the other senior leaders, and cultivate a healthy, collaborative culture. They are very comfortable identifying gaps in processes and managing highly complex, detailed work without getting overwhelmed.

What's a CTO?

At a small company or startup, "senior management" often implies that your title is CTO. And yet the CTO job is one of the least well-defined roles in the technical world. If you're a CTO, what are you supposed to be doing? If you want to become a CTO, what do you need to do to get there?

Let's start by talking about what a CTO is not. CTO is not an engineering role. CTO is not the top of the technical ladder, and it is not the natural progression engineers should strive to achieve over the course of their careers. It's not a role most people who love coding, architecture, and deep technical design would enjoy doing. It follows that the CTO is not necessarily the best engineer in the company.

Now, with that out of the way: what is a CTO, if not the best writer of code and the natural pinnacle of the engineering ladder?

The challenge in defining the role of the CTO is that if you look at the folks who hold that title, you'll see many different manifestations. Some are the technical cofounders. Others were the best of the early engineers. Some started at the company with the title, while others (such as myself) were promoted into it over time. Some became CTO after being the VP of Engineering. Some focus on the people and processes of engineering, hiring, and recruiting. Others focus on the technical architecture or the product roadmap. Some CTOs are the face of the company to the external technical world. Some CTOs have no direct reports, while others manage the entire engineering organization.

After looking at all these different examples, the best definition I can give you is that the CTO is the technical leader at the company's current stage of evolution. To me, that definition is rather unsatisfying, and misses the hardest part of the job. To expand, *the CTO should be the strategic technical executive the company needs in its current stage of evolution.*

What do I mean by "strategic"? The CTO thinks about the long term, and helps to plan the future of the business and the elements that make that possible.

What do I mean by "executive"? The CTO takes that strategic thinking and helps to make it real and operational by breaking down the problem and directing people to execute against it.

So, what does a CTO actually *do*?

First and foremost, a CTO must care about and understand the business, and be able to shape business strategy through the lens of technology. He is an executive first, a technologist second. If the CTO doesn't have a seat at the executive table and doesn't understand the business challenges the company faces, there's no way he can guide the technology to solve those challenges. The CTO may identify areas where technology can be used to create new or bigger lines of business for the company that align with the overall company strategies. Or he may simply ensure that the technology is always evolving to anticipate and enable the potential futures of the business and product roadmap.

No matter what, the CTO must understand where the biggest technical opportunities and risks for the business are and focus on capitalizing on them. If he is focused on recruiting, retention, process, and people management, that's because it's the most important thing for the technology team to focus on at the time. I present this idea in contrast to the notion that the CTO should focus on purely technical issues, as the "chief nerd."

Strong CTOs also have significant management responsibility and influence. This doesn't always mean they're deeply involved in the day-to-day management, but part of maintaining your ability to shape the direction of the business and the business strategy is putting people behind solving the problems you believe will impact the business. Other executives will have ideas and needs for technology. The CTO must protect the technology team from becoming a pure execution arm for ideas without tending to its own needs and its own ideas.

Things get tricky when the team grows to be very large, and the CTO starts to hire VPs to manage all the people. Many CTOs give up all of their management responsibilities to their VPs, sometimes going so far as to not even have the VPs reporting to them. It's incredibly difficult to maintain influence and effectiveness as an executive with no reporting power.

I saw this clearly at a previous employer, where the most senior members of the technical staff in large business areas would often hold the title of CTO for that area. These people were always highly respected and technically capable. They understood the business and its technical challenges, and were often called on to help inspire the engineering team and help with recruiting. Yet they had trouble being successful because they often lacked the direct management oversight of any teams, and because technology was frequently viewed as an execution arm, they didn't have much strategic influence.

If you're a leader with no power over business strategy and no ability to allocate people to important tasks, you're at best at the mercy of your influence with other executives and managers, and at worst a figurehead. You can't give up the responsibility of management without giving up the power that comes with it.

The CTO who doesn't also have the authority of management must be able to get things done purely by influencing the organization. If the managers won't actually give people and time to work on the areas that the CTO believes are important, he is rendered effectively powerless. If you give up management, you're giving up the most important power you ever had over the business strategy, and you effectively have nothing but your organizational goodwill and your own two hands.

My advice for aspiring CTOs is to remember that it's a business strategy job first and foremost. It's also a management job. If you don't care about the business your company is running—if you're not willing to take ultimate responsibility for having a large team of people effectively attacking that business—then CTO is not the job for you.

Ask the CTO: Where Do I Fit?

I'm so confused by the various titles in engineering leadership. CTO, VP of Engineering—what's the difference? How do I know which of these roles I want to hold?

I understand the confusion. There are numerous popular articles out there on the difference between these roles because it's hard to give specifics about them beyond "it depends." Of course, there are many different ways to do these two jobs.

To decide which job you want, you may ask yourself a few questions. Do you think you might cofound a company someday? Do you want to help oversee technical architecture and set up processes and guidelines for evolving it? Are you also willing to go deep into understanding the business side of things in order to ground that technical architecture in the company's growth? Are you willing to do external events, speaking, selling to customers, and recruiting senior managers and engineers? Are you willing to deal with managing and mentoring senior individual contributors? You might be a good CTO.

How about management? Do you enjoy managing people? Do you enjoy making engineering processes more efficient? Do you like to have a broad view of the work being done by a team and a hand in prioritizing that work? Are you fascinated by organizational structure? Are you good at partnering with product managers? Are you willing to exchange depth of focus into technology details for focus on the effectiveness of the overall team? Would you rather sit in a roadmap-planning meeting than an architecture review? You're probably more interested in following the VP of Engineering path.

Some people will have a mix of answers. I've been a VP of Engineering and a CTO, both times via promotion. I was always thinking about the technical architecture, but when my job required it I was happy to focus on the organization. For me, though, having only an organizational focus is not enough to keep me motivated. I like to think about organizational structure, but I get bored with the details of process and roadmap planning, and need to have some technical and business strategy oversight to be engaged.

The fastest route to becoming a CTO is to be a technical cofounder, but that guarantees you the job only so long as you and the startup grow together. The fastest route to being a VP of Engineering is to get management experience at a larger organization and then join a growing startup.

I'll leave you with some advice my own VP of Engineering once gave me: "Wanting to be a CTO (or VP of Engineering) is like wanting to be married. Remember that it's not just the title, it's also the company and the people that matter." Titles are definitely not everything.

Changing Priorities

One morning, the CEO wakes up and in the fresh morning light, she has a revelation. She sees an opportunity for the company to develop a new product line that can push the business to a new stage of growth. She spends some time sketching out this vision and presents it to the rest of the senior leadership team. On board with this change, they begin to make the moves needed to make the vision a reality. But it doesn't happen quickly. There are in-flight projects to worry about. Some things are almost complete, and it would be a shame to cancel them before they got finished. All of these concerns mean that the teams are slow to come together to work on the initiative, until suddenly the question comes down: *so, why aren't you working on the top priority?*

Priority changes from senior management can sometimes happen without warning. Leaders who are removed from the day-to-day schedules of the teams can forget that teams have long priority lists that may have been mapped out weeks or months ago and may take weeks or months to complete. So when these leaders see an opportunity or feel that the priorities of the organization need to change, they often expect that change to happen immediately, without consideration for the reality of the current state of affairs.

This question may be asked of managers at every level, but most often originates from senior management. Expect to get this question from your boss. When you feel the need to ask it of your own teams, ask yourself why they don't understand what the priorities are and what they should be cutting to address them.

Do you know what the top priority is? Do your teams know what it is? Do the developers on those teams know what it is? Sometimes the answer to this question is simply a matter of communication. You don't know what the top priority is, or you didn't communicate it clearly and urgently to your management team, and they didn't communicate it clearly and urgently to their development teams. You didn't explicitly go through the list of things in flight and kill or postpone work in order to make room for this priority. You need to do that, if it's truly urgent. Saying something is top priority is one thing, but making the actual tradeoffs on the schedule to get people moving on it is completely different.

We forget that the people above us or in different organizations don't have the same detailed understanding of what our teams are currently doing, and why. I don't believe it's necessary to constantly provide minute details to peers and your manager for every team in your large organization. However, when you're taken to task for not focusing on the right priority, it's a sign that you and the CEO have a misaligned understanding of reality, and you need to get on the same page. Your team may be crunching to stabilize a system that's causing frequent outages, or in the last push of a major project that has been ongoing for a long time. If you think that the team needs to finish their current work before shifting to the new top priority, you must communicate that clearly.

Be prepared to push both up and down to maintain or change focus. If you think a big project should be finished before slotting in new work, get as many details as possible about the value of that project, its current status, and the expected timeline. Be realistic. If someone above you has changed business focus so urgently that he's willing to have this conversation, expect that you'll probably need to compromise on the current work in progress and cut some parts short or move some people off of it. Your team may not be happy with the change. People generally don't enjoy being pulled off of what they're working on for a new executive whim, especially if they believe their current work is important.

The more senior the management and leadership position you take in a company, the more the job becomes making sure that the organization moves in the direction it needs to move in, and that includes changing direction when needed. You do this by clearly communicating the direction to your teams, and making sure they understand it and are taking the necessary steps to change course. Ask your teams for the list of projects the change will impact, so that you can communicate it upward. This will force your management team to actually think about the new initiative and start to plan for it. Ask for the goals of the initiative from its originator, and see how you can combine those goals with work already in flight.

Finally, never underestimate how many times and how many ways something needs to be said before it sinks in. Communication in a large organization is hard. In my experience, most people need to hear something at least three times before it really sinks in. You're going to tell your own senior management and leadership team. You'll hold an all hands meeting. You may need to send some email detailing the changes as well. A little bit of communication planning can go a long way in such situations. Try to anticipate the questions you might get and prepare answers for those questions. Be as clear as possible about the projects or structure to be changed, so there's little room for confusion. And don't forget to sell this change as a good thing!

You'll also need to repeat information when you're communicating up. When you want your boss to act on something, expect that you'll need to tell him the same thing three times before he actually listens. The first two times, the issue may still resolve itself, but the third time, it's a sign that something bigger needs to happen. You may be surprised to find that you start acting the same way toward your team. Many problems will get raised to you and then resolved on their own, so you may decide you need a degree of sustained struggle from your team before it's time to step in. I'm not recommending that you adopt the "three-times rule" as a policy, but it does tend to happen, whether you plan for it or not.

The larger the organization, the harder it'll be to change priorities quickly. If you're working for a growing startup with a founder CEO, this slowness will frustrate him. The best thing you can do to manage this situation is to proactively keep the CEO informed about what's happening and why. Do your best to show that you understand his priorities and tell him about the concrete steps you're taking to meet those priorities.

Setting the Strategy

In the summer of 2014, as the SVP of Engineering at Rent the Runway, I had a big challenge ahead of me. My CEO told me that she wanted to put me up for promotion to CTO at the next board meeting, but that as part of that promotion, I'd be asked to present the technology strategy to the board. She then returned every attempt I made at giving her this strategy until I finally created something that met her standards. And the rest, as they say, is history.

I suspect she didn't have to put me through this exercise to promote me. The board was very happy with the fact that I had grown the team and developed the technology to a point of stability and high feature throughput. I'm incredibly grateful that she did push me through this exercise, however. During that process, I went from having only a vague idea of what setting strategy meant to having a concrete, forward-thinking strategy encompassing a way to think about both the technical architecture and the engineering team's structure, which in turn ultimately influenced the way the company itself thought about its overall structure.

When I talk about senior leadership, I emphasize strategy as a critical element. Most people don't even really know where to begin when it comes to setting strategy at the senior level. I know I didn't. I had coaching, from the CEO and from an executive CTO coach that I was working with. I solicited input from my peers on the executive team. I posed theoretical questions to the senior members of the engineering team and used them to help me see some of the detailed problems. I certainly didn't do it alone. So, with that in mind, what does setting technology strategy look like?

DO A LOT OF RESEARCH

I started by considering the team, the technology we had currently built, and the company. I asked the engineering team where their pain points were. I asked several executives in various areas where they expected growth to come from in the future. Then, I asked myself several questions. I considered where the scaling challenges were now, and where they might be in the future. I examined the engineering team and found its productivity bottlenecks. I studied the technology landscape and wondered how it might change in the near future, especially as it pertained to personalization and mobile development.

COMBINE YOUR RESEARCH AND YOUR IDEAS

Armed with my conclusions about the existing systems, teams, and bottlenecks, and having imagined the places where we could make things more efficient, expand features, or improve the business, I used the data to come up with a rough idea for a possible future. I spent some time sitting alone in a room with a whiteboard or paper and drawing out the systems in place at our company, slicing and dicing the systems and teams across various common attributes. For example, I looked at systems that were customer-facing versus systems that were internal operations–facing (such as warehouse and customer service tools); I looked at backend versus frontend. Because our technology had to model most of the business data to operate, I realized I had a unique insight into the way that data flowed and changed, and possible axes for evolution.

DRAFT A STRATEGY

Once I had that data mapped, I could plan out actionable ideas to improve operational efficiency, expand features, and grow the business. I considered the places we would want to potentially limit or expand information sharing between systems. Did we want personalization systems that tried to operate on the real-time state of the world at all times, or did we want personalization systems that acted more like a view adjustment on subsets of the data? How could we use various product and operational attributes in parts of the data flow to have operational as well as personalized input into what our customers experience? All of this thinking forced me to consider the structure of the business, the needs of the customer—both internal and external—and possible future evolutions. By doing this research and speculation, I was able to design a technology strategy that supported these factors into the future.

CONSIDER YOUR BOARD'S COMMUNICATION STYLE

I said earlier that the CEO sent back many of my attempts to get this work done. Really, she rejected two things. The first was an underdeveloped strategic plan that was almost entirely about system and architectural details and offered very few forward-thinking ideas beyond the next 6 to 12 months. It certainly did not attempt to address the business drivers that were critical to the team's success. The second was my slide deck. As a speaker, I've been trained to make slide decks that are sparse, in support of an audience that listens closely. This board needed a deck that was very dense with information. It's not uncommon for company boards to read through the slide deck before a meeting, so that the meetings can be focused more on details than on presentations. I didn't understand this at the time, so I wasted a lot of energy trying to make something that wasn't informative on paper. Lesson learned.

As you can see from this tale, good technology strategy here meant several things. It meant technology architecture, yes. It also meant team structure. It meant understanding the underpinnings of the business and the directions in which it was headed. I like to describe technology strategy for product-focused companies as something that "enables the many potential futures of the business." It's not just a reactive document that tries to account for current problems, but it anticipates and enables future growth. If you're in a product-focused business, this is the heart of your technology strategy. It's not about actually deciding the product's direction, but about enabling the larger roadmap to play out successfully.

The hardest part, in many ways, was getting started. The second-hardest part was getting comfortable making a guess about the future with highly imperfect information. Going through this exercise was the difference between my ability to lead in a reactive fashion, looking at the known environment and making plans to accommodate it, and my ability to lead in a forward-thinking fashion. Now I had an idea about where we needed to go, as an architecture, as a team, and as a company.

After I clarified this architecture for myself, leading became in many ways much easier. I could show the engineering team a vision of where we would go as a technology platform, not just what the product roadmap looked like. I had ideas about things we could work on that would directly move the company forward, beyond making the technology work. The architecture led the technology's organizational strategy, which ended up leading some of the company's organizational strategy—something I was quite proud of being able to influence.

Challenging Situations: Delivering Bad News

We all have times when we have to deliver bad news to our teams. Maybe the company is having layoffs. Maybe the team is being disbanded in order to give more support to other projects, and everyone is being scattered onto other teams. Perhaps there's a policy change that you know will be unpopular. Those roadmap changes we just talked about sometimes fall into this category as well. You, the manager, have to be the messenger and deliver this news, but you know the team isn't going to be happy.

What do you do in this situation? Well, communication is key. As a person in senior leadership, you'll need to excel at communicating sensitive information to large groups. Here are a few dos and don'ts:

- **Don't blast an impersonal message to a large group.** The worst way to communicate bad news is via impersonal mediums like email and chat, especially mediums with commenting abilities. Your team deserves to hear the message coming from your mouth directly, and without you to guide the message, you can expect some misunderstandings and bubbling animosity. That being said, the second-worst way to deliver this message, especially to a large group that you know won't be happy, is with them all in a room at once. You may think that trying to communicate bad news to everyone at once is the best way to keep it from spreading before everyone has heard it, but the result is still impersonal. It's hard to see everyone's reaction, and one or two deeply unhappy members can quickly rile up the whole team before the message has had the chance to sink in.

- **Do talk to individuals as much as possible.** Instead of impersonal or group-based communication, try your best to talk to people individually about the news. Think about the people who are going to have the strongest reaction, and try to tailor the news to them. Give them space one-on-one to react, to ask questions, to get it straight from you. And as necessary, make it clear that these are the marching orders and that you need your people to be on board with the changes, even if they don't love them. In a case where you need to get the message out to your whole organization, talk to your managers first, give them talking points, and then let them share with their teams before bringing the whole group together.

- **Don't force yourself to deliver a message you can't stand behind.** You may have a hard time delivering this news because you don't like it yourself. Maybe you violently disagree with the policy change. Maybe you hate the fact that the team is being split up. If you absolutely can't deliver the news in a way that won't betray your strong disagreement, you might need to have someone else help you deliver it. Perhaps you ask another executive to step in, or maybe someone from HR. Depending on the size of your team, you can deliver the information to a trusted lieutenant and have that person help to share it. As someone in senior leadership, you have to learn how to maturely handle decisions you don't agree with, but that doesn't mean you have to go it alone.

- **Do be honest about the likely outcomes.** The more you can commit to the direction specified by the news yourself, the easier this will be. If there are layoffs, acknowledge that this process is not fun but that everyone needs the company to survive. If a team is being disbanded, feel free to point out both the accomplishments of the team up to this point and the changes that make this the right path forward, and emphasize that there are many new opportunities now for learning and growth. Being forthright with people will help them trust you more and make them more likely to tolerate the unhappy news well.

- **Do think about how you would like to be told.** One piece of news that you may have to deliver someday is the news of your leaving. In fact, you've probably had the experience of resigning from a job already, or moving from one team to another. How did you communicate that news? Did you send a memo? Well, maybe to HR, but to the rest of the team you probably pulled people aside to tell them face-to-face if you thought they would want to know before it was public. You may have had a going-away party, written a farewell letter, or given a final lecture to the team about what you learned in your time at the company. It's OK in some circumstances to celebrate these sad changes, so long as you can do it with grace. All of these lessons apply to delivering hard news to your team.

Ask the CTO: I Have a Nontechnical Boss

This is the first time I've ever had a boss who was nontechnical, and it's turning out to be really hard. How do I effectively manage this relationship?

The first time I ever experienced a fully nontechnical manager was when I started reporting to the CEO at Rent the Runway. Reporting to a nontechnical manager can be a total culture clash. Fortunately, a few best practices will help you manage this relationship:

- **Don't hide information behind jargon, and be careful with details.** Your new boss is probably very smart, but he may not have the patience for technical jargon, and it is very unlikely that he wants to hear a ton of details about nuanced technical decisions. Part of this process is learning to distinguish the type of information that is valuable to communicate from the type that is not.

- **Expect that you will need to run your 1-1s with your new boss, so come prepared with a list of topics.** Busy executives can be frustratingly hard to pin down; even getting 1-1 time is a challenge, so don't waste the time you have. Up until now you may have expected both parties to bring some specific topics to the meeting, but that won't necessarily be the case now, so always come prepared. If you're having trouble getting your 1-1 time honored, send the agenda in advance to remind your boss that you need her attention—and it never hurts to be on good terms with the executive assistant who manages her calendar!

- **Try to bring solutions, not problems to be solved.** CEOs generally do not want to hear about how things are failing, nor do they want to hear about your disagreements with your peers or your troubles managing. In the case that you have a CEO who doesn't want to hear too much about problems, respect that you're not going to get much coaching from him on the management side of things, and find another person to get that from. With that said, don't shy away from delivering bad news.

- **Ask for advice.** It sounds contrary to my advice to bring solutions, but nothing shows respect like asking for someone's advice. Your boss may not want to be stuck solving your problems, but you can bet that she'll be happy to provide feedback if you phrase it as needing advice.

- **Don't be afraid to repeat yourself.** If you've brought up an important issue that seems to have been forgotten, bring it up again if it's really important. You may have to do this a few times before you get any traction. Three times is often the magic number.

- **Be supportive.** Always ask if there is more you can be doing to help. As much as you can, show that you are there to support your boss and the company.

- **Actively look for coaching and skill development in other places.** You no longer have a manager; you have a boss. You probably still need to work on some skill development the first time you work in a senior leadership position, so get a coach, ask for training, and create a peer group outside of the company to support you through this brave new world.

Senior Peers in Other Functions

Many of the major revelations for me as I moved into senior management involved the relationships that I needed to develop with the other people on the executive and leadership teams. I had the opportunity to work with people I respected across many functions, and because we had a large and diverse leadership team I got to know many different types of senior leaders. I got along better with some than with others, and I evolved my perspective by learning from both types of interactions.

Senior leaders, more than any other group in a company, must actively practice first-team focus (introduced in Chapter 6). They are dedicated first and foremost to the business and its success, and secondly to the success of their departments as a way of contributing to the overall business success. Leadership books such as Patrick Lencioni's *The Five Dysfunctions of a Team*[2] write about this relationship, and while many of us start to practice this type of leadership with other engineering managers as our "first team," senior leadership is often the first time your peers all operate in a very different way than you're used to, and having few to no peers on your first team who are fellow engineers can feel very isolating.

So what does it feel like to work well with cross-functional peers? To start with, you let them own their areas, and they let you own yours. Many of us learn how to do this earlier in our careers, when we have to work with senior designers, product managers, or other business team members, but if you haven't learned how to let a peer own her specialty, now is the time. Giving her respectful deference when it comes to her turf is fundamental. If you disagree with her management style or application of her skill set in places where it isn't directly

---

2 Patrick Lencioni, *The Five Dysfunctions of a Team: A Leadership Fable* (San Francisco: Jossey-Bass, 2002).

affecting your team, you treat that disagreement like you would treat a good friend who happens to date people you don't love. Unless she asks for your advice, try to stay out of it as much as possible, and certainly approach any disagreement you choose to discuss with kindness. Be willing to let those differences lie.

Of course, you'll disagree with your peers. The place for this disagreement is either one-on-one or in your leadership team meetings. These meetings are where you hash out differences of opinion on strategy, challenges the company is facing, and details of direction setting. If the numbers don't make sense, ask the CFO in the context of these meetings. Also expect to defend your own technical decisions and roadmap in this setting.

This brings us to the second element of trust, beyond trusting someone's abilities. In *The Five Dysfunctions of a Team*, Lencioni notes that absence of trust is a fundamental dysfunction. In this case, what's missing is the trust that your peers are actively trying to do their best for the organization, that they are not trying to manipulate situations, undermine you, or otherwise get their own way. So, beyond establishing basic respect for your peers' ownership of their turf and respect for their abilities as functional experts, you also have to put aside the idea that they're acting in irrational or self-interested ways when they disagree with you or do things you don't like.

Establishing this fundamental trust is really difficult. You will probably have some degree of a cultural clash with some, if not all, of your peers. The values that make a great CTO are probably slightly different from those that make a great CFO, CMO, VP of Operations, and so on. A very common clash occurs between people who are extremely analytically driven and those who are more creatively or intuitively focused. Another is between the people who prefer to embrace agility and change (and, yes, sometimes disorder) and those who push for more long-term planning, deadlines, and budgets. You have to figure out how to understand and trust everyone's styles across the spectrum.

Engineers frequently struggle with the transition to respecting and communicating well with diverse peers. I believe the struggle with respect is a side effect of the current tech culture, which tells us that engineers are the smartest people in the room. It can't be said strongly enough: your peers who are not analytically driven are not stupid. On the flip side, we undermine ourselves when we fail to talk so that nontechnical peers can understand what we're saying. Throwing out jargon to people who aren't familiar with it—and who don't even need to be familiar with it—makes *us* look stupid to *them*. We therefore need to figure out how to communicate the complexity of our work in a way that an intelligent but nontechnical peer can understand.

The final element of this first-team trust and respect is the cone of silence. Disagreements that happen in the context of the leadership team don't exist to the wider team. Once a decision is made, we commit to that decision and put on a united front in front of our engineering teams and anyone else in the company. It's easier said than done—I've struggled frequently with hiding my own disagreements with my peers. Letting go when you don't get your way, especially when you don't feel that your objections have been heard, is hard, and it will have to happen from time to time. At this level especially, you must decide whether you want to fall in line or quit. The middle ground, openly disagreeing with your peers, does nothing but make the situation worse for everyone.

The Echo

As the senior-most person in an organization, you'll be watched more closely than you've ever been watched in your life. Your presence causes people to focus all of their attention on you. They seek out your approval and try to avoid your criticism. Shifting your mindset away from "one of the team" to "the person in charge," especially if you came up through the team and grew the team yourself, is a challenge for many at this level.

You're no longer one of the team. Your first team is comprised of your peers at the leadership/executive level, and your reporting structure has now become your second team. If you shift your mindset successfully, you will probably start to detach socially a little bit from the overall organization. When there's a happy hour, you go for a drink and then leave the team to socialize. Closing down the bar with your whole organization will tend to have bad consequences for everyone, so I strongly advise that you avoid doing that with any regularity. Socializing heavily with your team outside of working hours is a thing of the past.

You need to detach for a few reasons. First, if you don't detach, you're likely to be accused of playing favorites. In fact, you probably *will* play favorites if you maintain very strong social ties with people who report up to you on the team. This hurts, but it is true. Maybe you don't care, but personally I found that having my team believe I was playing favorites made the overall team unhappy and made my job a lot harder.

Second, you need to detach because you need to learn how to lead effectively, and leading effectively requires people to take your words seriously. The downside of leading at this level is that with a throwaway comment, you can cause people to change their whole focus. This is bad, unless you're aware of that and actually make use of it appropriately. If you try to maintain a "buddy" image, your reports are going to have a hard time distinguishing between their buddy thinking out loud and their boss asking them to focus on something.

Detaching also means being thoughtful about where you spend your time. As the senior leader, you'll often suck all of the oxygen out of a room. Your mere presence will change the tone and structure of meetings you attend. If you aren't careful, you'll end up pontificating and change the direction of a project because you had a great brainstorm in a one-off meeting you decided to drop into. It sucks! I know! It's frustrating that you can no longer be one of the team whose ideas are there to be evaluated and potentially rejected—but you are no longer that person.

If you've ever worked with anyone who overlapped with Steve Jobs at Apple, chances are you've heard that person talk about "Steve" and the impact he had on some project he or she was working on. Apple employees used the specter of Steve to argue for and against decisions, as a moral compass for what the organization should be doing. The culture you build and reinforce will have some of this effect on your company. They may not refer to you by name, but as you choose which behaviors to model in front of the team, they will learn those behaviors and copy them. If you yell, they learn that yelling is OK. If you openly make mistakes and apologize, they learn that it's OK to make mistakes. If you always ask the same set of questions about a project, people start to ask those questions themselves. If you value certain roles and responsibilities openly above other roles, ambitious people will seek out those valued positions. Use this power for good.

There are other reasons you need to detach. You're going to be part of hard decisions that will impact the whole business, and these decisions may cause you a great deal of stress. It won't be appropriate to discuss these decisions with other people at the company. It's deeply tempting to rant to those people you consider friends in your reporting team about the challenges of your position, but this is a bad idea. As their leader, you can easily undermine their confidence by sharing worries that they can't do anything to mitigate. Transparency that may have been harmless or even possibly helpful at lower levels of management can become incredibly damaging to the stability of your team at this level.

You may not be "one of the team," but that doesn't mean you should stop caring about the team as individuals. In fact, I encourage you to care *more* about people as individuals, at least in a small way. Taking the time to get to know as many people as you can as humans—asking them about their families or hobbies or interests—is a good way to help them feel that they're part of a group that cares about them.

As you grow more detached from the team, it can be easy to start to dehumanize people and treat them like cogs. People can tell when that's happening, when their leaders stop caring about the individuals in the organization. They're less likely to feel committed to giving their all, to taking risks and pushing through hard circumstances, if they feel that no one really cares about them personally. Nurturing that kind of connection, even at a superficial-seeming level, helps to reinforce that you *do* care about them and not just their current projects or work output; that you know they're people outside of work. It grounds you without attaching you too strongly to individuals. You'll have to make hard calls as an executive, but your team deserves a leader who's able to be kind even while making those hard calls.

You're a role model. What kind of leaders do you want to develop? What kind of legacy do you want to leave?

Ruling with Fear, Guiding with Trust

Camille considers herself to be a good leader: technical, charismatic, capable of making decisions and getting things done. She's also sometimes short-tempered, and when people don't live up to her expectations or things go wrong, she can be visibly annoyed and openly angry. She doesn't realize that this hard edge and short temper are making people afraid of her. They don't want to risk getting blamed for failure or openly criticized for making a mistake, so they take fewer risks and hide their mistakes. Camille has accidentally created a culture of fear.

Michael is also a good leader: technical, charismatic, capable of making decisions and getting things done. He's also good at keeping his cool. Instead of getting tense and angry, he gets curious when things don't seem to be going well. His first instinct is to ask questions, and these questions often cause the team to come to their own realizations about what's going wrong.

You might be surprised to learn that the story you just read is true: it's my own experience accidentally creating a culture of fear when I became a senior leader. Here's an excerpt from my very first review as a senior leader:

> *Even those who love you on your team admit being fearful of you and your potential criticism. People are afraid to take risks or fail in front of you because they are scared of being publicly reprimanded in front of their*

peers. What your attacks have done is create a culture where members of the team are afraid to engage with you, to ask you questions, or to ask you for feedback—which then leads to a vicious cycle of you not trusting them and them making mistakes.

As you can imagine, this was a shocking and uncomfortable thing to hear. And while I can make dozens of excuses for it—people take criticism more harshly from women, I came from the finance industry where this was a normal culture, everyone just needs to toughen up—it was clearly a problem. People were afraid to take risks, and if you want to have an independent team capable of setting their own direction and pushing themselves, you need them to take risks.

How do you know if you're creating a culture of fear? It can come from placing a high value on being correct and following the rules, and having a strong affinity for hierarchy-based leadership. I also believe that coming from places where conflict was openly tolerated, if not actively encouraged, made me even more likely to create this culture. Engineering culture has a high tolerance for open debate to resolve conflict, so leaders who come from heavily engineering-focused backgrounds may feel particularly comfortable aggressively sparring with others over issues. Unfortunately, when you're the leader, the dynamic changes, and those who may have fought back when you were an individual contributor will feel threatened by you as a leader.

CORRECTING A CULTURE OF FEAR

- **Practice relatedness.** One marker of the culture of fear is a tendency to treat people impersonally. In my early days of management I had a habit of focusing too much on efficiency. I wanted to dive right into the issue at hand, to get into the intellectual discussion, the status update, the problem that needed to be solved. I didn't take a lot of time for small talk, for getting to know my team as people or letting them see me as a person, and as a result, I didn't have much of a personal relationship with them.

  If you want a team that feels comfortable taking risks and making mistakes, one of the core requirements is a sense of belonging and safety. This means you need to take a little time for small talk. Ask people about themselves, get to know them as humans, let them know you. Most people are scared to take risks in front of people they think will reject them if they fail. Intentionally or not, by neglecting to create even basic personal relationships with many of my team members, I made them afraid of how I would react to mistakes, questions, and failures.

- **Apologize.** When you screw up, apologize. Practice apologizing honestly and briefly.

    "I'm sorry, I should not have yelled at you and I have no excuse for my bad behavior."

    "I'm sorry, I did not listen to you and I know I contributed to your frustration at this situation."

    "I'm sorry, I made a mistake when I neglected to tell you about Bob."

    Apologies don't need to be drawn-out affairs. A short apology that takes responsibility for your role in creating a negative situation or hurting another person is all that is necessary. If you go too long, it often turns into an excuse or a distraction. The goal with apologizing is to show people that you know your behavior has an impact on others, and to role-model for them that it's OK to make a mistake but that you should apologize when you hurt other people. You're showing the team that apologizing doesn't make you weaker—it makes the whole team stronger.

- **Get curious.** When you disagree with something, stop to ask why. Not every disagreement is an undermining of your authority. When you take the time to seek out more information about something with which you disagree, you'll often find that you were reacting to something you didn't really understand. For those of us who care about doing the right thing or making the best decision, attacking something we disagree with makes that harder. When we attack, many people evade or shut down, and they learn that it's a good idea to hide information from us so we won't attack or criticize them. When you get curious and learn how to turn that disagreement into honest questioning, you can learn more about other perspectives on the issue because your team will open up. This is how you get the most information out and help everyone make the best decisions.

- **Learn how to hold people accountable without making them bad.** As a leader, you want your teams to do their jobs well. If they fail to meet their responsibilities, you'll be the person to hold them accountable. But it doesn't start with responsibility and end with consequences. Other elements are needed along the way. How do you measure success? Does the team have the capabilities needed to succeed? Are you providing feedback along the way? I think many leaders forget these requirements and hope they can get a junior team to achieve something just by setting the goal clearly, or believe a more experienced team shouldn't ever need feedback. Think of the times you've made a person or a team out to be "bad" because they failed. Are you holding yourself accountable for setting them up for success? When everything is clear and you've all done your best, I bet you'll find that accountability comes with far less character judgment, because you all clearly see what has happened.

The culture of fear is pretty common in technology, and it survives best in environments where things are otherwise going well. Don't be fooled by external circumstances that enable your bad behavior. If you're feared but respected, the company is growing, and the team is working on interesting problems, you might get along OK for a time. However, if you lose any of these elements, you can expect to see people who have better options leave for greener pastures. I know firsthand that having a team that fears but respects you isn't enough when they're frustrated by other things happening around them. So work on softening your rough edges, practice caring about your team as humans, and get curious. Building a culture of trust takes time, but the results are well worth it.

True North

A core role of senior leadership is sometimes overlooked. This role, played by the senior leader of a functional area (the CTO plays it for technology, the CFO plays it for finance, etc.), sets the baseline of what excellence looks like in this function. I call it "True North."

True North represents the core principles that a person in a functional role must keep in mind as he does his job. For a product leader, True North includes thinking of the users and their needs first and foremost, measuring and experimenting as much as possible, and pushing back on projects that don't address the stated goals of a team. For a CFO, True North includes looking at the numbers, at the costs of work and at the potential value, and making sure that you've considered how to make those numbers work in your company's favor, that the company isn't accidentally spending more money than expected, that the team knows when it's at risk for going over budget.

For a technical leader, True North means making sure that you've done your job getting things ready to go into production. It means you have honored your agreed-upon policies for review, operational oversight, and testing. It means that you won't put something into production that you don't believe is ready for your users to experience. It means you're creating software and systems you're proud of.

Technology leaders must help set the standard for True North in their organizations for different types of projects and exposures. Another way to think of this is through the lens of risk analysis. Risk analysis doesn't mean that we don't take risks. Some things that are generally considered "bad" can be OK under certain circumstances. These include:

- Having a single point of failure
- Having known bugs and issues
- Being unable to tolerate high load
- Losing data
- Putting out code that is undertested
- Having slow performance

There are situations and companies in which all of those risks are acceptable to take. That being said, True North helps us understand that all these issues must be carefully considered when we put code into production. Just because these rules have exceptions doesn't mean we forget that they exist.

I call this concept True North because it's important to understand it as an underlying pull, as a guiding instinct that we as leaders have developed over time and strive to help our teams as a whole develop as well. When our teams develop this instinct, they can be trusted to independently follow these guidelines without much direction or nudging.

The True North for each functional area is slightly different, so there will be natural tension in the organization. Product managers may care more about the user experience and less about the production support burden. The finance team may care more about the overall cost of infrastructure and less about the risks for availability. This tension is healthy because it forces us to reckon with all of the risks, not just the risks that our particular function cares about.

When you examine your role as a leader, looking at the ways that you set True North can help you understand your strengths and areas of ownership. If you consider yourself a technical leader, part of your job should be setting True North for large facets of the critical technology. As a CTO for a commerce company, I set True North for most fundamental technical decision making around production readiness, scaling, systems design, architecture, testing, and language choice. That doesn't mean I made all of those decisions, but I guided the standards by which such decisions would be evaluated. I delegated True North for matters of mobile and UI-specific development, but pushed the senior technical staff in those areas to articulate what the standards looked like.

True North leaders rely on the wisdom they've developed over time to make fast decisions when they don't have time to delve into all the details. If you want to become this type of leader, you *must* spend enough time early in your career to hone these instincts in order to be comfortable making fast judgment calls. That means staying technical; following through with projects, languages, or frameworks long enough to learn more than their basics; and also pushing yourself to keep learning new things even when your day-to-day doesn't involve writing code.

Recommended Reading

- Arbinger Institute, *Leadership and Self-Deception: Getting Out of the Box* (San Francisco: Berrett-Koehler, 2000).

- Brené Brown, *Daring Greatly: How the Courage to Be Vulnerable Transforms the Way We Live, Love, Parent, and Lead* (New York: Gotham Books, 2012).

- Peter F. Drucker, *The Effective Executive* (New York: HarperBusiness Essentials, 2002).

- Marshall Goldsmith and Mark Reiter, *What Got You Here Won't Get You There: How Successful People Become Even More Successful* (New York: Hyperion, 2007).

- Andrew S. Grove, *High Output Management* (New York: Vintage Books, 1983).

- L. David Marquet, *Turn the Ship Around! A True Story of Turning Followers into Leaders* (New York: Portfolio, 2012).

Assessing Your Own Experience

- At this level, your coaching and mentoring are likely to come from people outside of your company. You no longer have a manager, you have a boss. Do you have a professional coach, either provided by work or paid for yourself? This is a good investment even if your job doesn't pay for it. A coach can give you guidance and direct feedback, and unlike your friends, she's paid to listen to you talk.

- Beyond a coach, how is your support network of peers outside of your company? Do you know other senior managers at companies in your area? A peer group helps you see what the job looks like at other companies, and is a place where you can share experiences and get advice.

- Do you particularly admire any technology senior managers? What is it about them that you admire? What could you be doing to be more like them, if anything?

- Think back to the last time you needed to change priorities for part or all of your team. How did it go? What went well, and what didn't go well? How did you communicate the change to your team, and what was their reaction? If you were to do it again, what is one thing you would do differently?

- How well do you understand where your business is going for the foreseeable future? Do you understand the technology strategy that will help you get there? What are the critical areas of focus for the team, such as feature velocity, performance, technical innovation, and hiring, that need to evolve to reach the goals of the company? Where are the bottlenecks or opportunities for technology evolution to push the business forward?

- How is your relationship with the other members of the company's senior leadership team? Which relationships are good and which are bad? What could you do to improve the bad relationships? How well do you understand the priorities of these team members, and how well do you think they understand your priorities?

- If I asked your team which executives you got along with and which ones you hated, would they be able to tell me without hesitation? When the CEO or the leadership team comes to a decision that you don't agree with, are you capable of leaving that disagreement behind and supporting the decision to the rest of the company?

- Are you behaving like a role model for your team? Would you be happy to learn that people were emulating your behavior on a daily basis? When you sit in on meetings with your team, do you dominate the conversation or are you more interested in listening and observing?

- When was the last time you asked someone you don't talk to regularly about his life outside work? The last time someone emailed to say she was sick and couldn't come in, did you take a minute to wish her a quick recovery?

- What are the fundamental principles you want your senior engineers to be considering when they evaluate work and make decisions? Or, if you are focused more on organization than technology, what are the fundamental management principles you want your managers to be following when they lead their teams?

Bootstrapping Culture

When you are in the role of senior engineering leader, part of your job is to set the culture of your function. A common failing of first-time CTOs is to underestimate the importance of being clear and thoughtful about the culture of the engineering team. Whether you are growing a new team or reforming an existing team, neglecting the team culture is a sure-fire way to make your job harder. As the team grows and evolves, it's important to attend to your culture as you would attend to any other important piece of infrastructure that you rely on.

At Rent the Runway, I had the opportunity to set up many of the cultural elements of the engineering team. Because the team was still running on the classic, unstructured "scrappy startup" model when I joined, I was able to introduce many cultural structures and practices to both the team and its members. This process was a great learning experience for me.

For many people who are attracted to startup culture, the ideas of "structure" and "process" are seen as pointless at best and harmful at worst. I have seen surveys of startup teams in which the idea of introducing structure evoked such reactions as "slow" and "innovation-crushing." These respondents believed structure is the reason large companies move slowly, foster bureaucracy, and are generally boring places for bright people to work.

When talking about structure with skeptics, I try to reframe the discussion. Instead of talking about structure, I talk about learning. Instead of talking about process, I talk about transparency. We don't set up systems because structure and process have inherent value. We do it because we want to learn from our successes and our mistakes, and to share those successes and encode the lessons we learn from failures in a transparent way. This learning and sharing is how organizations become more stable and more scalable over time.

I'm hoping to help you develop a personal philosophy on company culture in addition to giving you ways to set up process and structure. If you want to create healthy teams, you need to have a sense of what is important to you, to your company, and to your growing group of colleagues. Consider not only what you care about, but also how you can scale that knowledge and effort effectively as the company and team grows and evolves. You're going to be trying out structures and processes and learning from them, but it's hard to learn if you don't have a basic theory to test, and you don't set out to prove or disprove hypotheses about that theory. So let's approach this culture-creating exercise scientifically, and see how you can think about the pieces of culture you might need in a logical fashion.

Early startups attract people who are capable of dealing with extremely high amounts of uncertainty and risk in exchange for equally high degrees of freedom to operate. There is no long-term guarantee that the company will succeed or even continue to exist for very long, no matter how strong the idea seems on paper. Often the market is unproven. Some signs look good and other signs look bad. There may be fierce competition from other companies, big and small. Furthermore, there is very little established work to build on. The code is unwritten. The business rules are not set up. It is hard to overstate how many decisions need to be made in the context of a startup, even one that has been growing for a couple of years. Everything from deciding on technology frameworks to deciding on office decorations is up for grabs.

Many of these initial decisions will be undone a couple of times before they settle. It's easy to think about changing a framework that didn't scale well with the company's technology needs, but things like vacation policy, core office hours, and even company values could change and evolve in a startup's first few years.

The most important thing for leaders to be willing to do in those early days—and *leaders* generally includes everyone in the company, not just the founders or executives—is to pick a strategy and run with it. Cultivate decisiveness in the face of a massive number of options. You have a problem? Figure out a solution and fix it. That solution doesn't work? Try something else. You don't need to find the perfect solution; you need to find something that will get you through to the next milestone, whether that milestone is the next release, the next growth spurt, the next funding round, or the next hire.

Sometimes companies decide to limit the decisions themselves, as in an organization that foregoes titles. Having no titles is in one sense a decision, but in another sense it's a decision that means you never need to decide what someone's title will be, you won't need to worry about promoting people to new titles, and you don't need to build up the apparatus that will make future decisions about titles because you have removed that as an option. Deciding not to decide right now is a popular option for new companies, because it really doesn't matter at the scale of a few people.

One of the greatest writings about organizational politics is a piece called "The Tyranny of Structurelessness" (*http://www.jofreeman.com/joreen/tyranny.htm*) by Jo Freeman. While the article is about early feminist/anarchist collectives, Freeman's insights apply equally well to startup culture. Pretending to lack structure tends to create hidden power structures resulting from the nature of human communication and the challenges of trying to scale that communication. Interestingly, Freeman describes a set of circumstances in which the unstructured group can, in fact, work:

1. **It is task oriented.** *Its function is very narrow and very specific, like putting on a conference or putting out a newspaper. It is the task that basically structures the group. The task determines what needs to be done and when it needs to be done. It provides a guide by which people can judge their actions and make plans for future activity.*

2. **It is relatively small and homogeneous.** *Homogeneity is necessary to insure that participants have a "common language" for interaction. People from widely different backgrounds may provide richness to a consciousness-raising group where each can learn from the others' experience, but too great a diversity among members of a task-oriented group means only that they continually misunderstand each other. Such diverse people interpret words and actions differently. They have different expectations about each other's behavior and judge the results according to different criteria. If everyone knows everyone else well enough to understand the nuances, they can be accommodated. Usually, they only lead to confusion and endless hours spent straightening out conflicts no one ever thought would arise.*

3. **There is a high degree of communication.** *Information must be passed on to everyone, opinions checked, work divided up, and participation assured in the relevant decisions. This is only possible if the group is small and people practically live together for the most crucial phases of the task. Needless to say, the number of*

interactions necessary to involve everybody increases geometrically with the number of participants. This inevitably limits group participants to about five, or excludes some from some of the decisions. Successful groups can be as large as 10 or 15, but only when they are in fact composed of several smaller subgroups which perform specific parts of the task, and whose members overlap with each other so that knowledge of what the different subgroups are doing can be passed around easily.

4. There is a low degree of skill specialization. *Not everyone has to be able to do everything, but everything must be able to be done by more than one person. Thus no one is indispensable. To a certain extent, people become interchangeable parts.*

Here Freeman describes a common scenario for many early-stage startups. Even when the overall company grows beyond the small group, the engineering team often pushes itself to stay unstructured. Hiring "full stack" engineers who are exclusively sourced from the professional and social networks of the current team results in low skill specialization and high homogeneity. Forcing the team to be collocated lowers communication barriers. And perhaps most critically, having an engineering team that operates solely as the execution arm of the product or founder makes the team highly task-oriented.

I will hazard a guess that some folks may bristle at this characterization of the common startup technology organization. After all, these engineering teams are often the well-paid darlings of the company! Be that as it may, the unstructured organization either displays characteristics that ultimately make it less self-directed than the members might wish to believe, or is run by hidden hierarchies and power dynamics. In many cases both things are true to some extent.

The example of the structureless team also applies to technical decisions and processes. There is a reason that you often find a lot of spaghetti code in early startups. When work is done to satisfy an immediate task, in a unified code base worked on by a team of interchangeables, the result is not usually a larger thoughtful structure, but a tweak here, a hack there—anything to get things done and moving forward. It's no surprise that we usually end up refactoring spaghetti code when we want to make it scalable, because refactoring usually involves identifying and explicitly drawing out structure in order to make the code base easier to read and work in.

That, in short, is the value of structure. Structure is how we scale, diversify, and take on more complex long-term tasks. We do it to our software, we do it to our teams, and we do it to our processes. In the same way that strong technical systems designers are capable of identifying and shaping underlying system structures, strong leaders are capable of identifying and shaping underlying team structures and dynamics, and doing so in a way that supports the long-term goals of the team and equips the individuals to achieve their best.

Nothing is more ridiculous than a small team with a rigid hierarchy. We would all think that a team of five people where the fifth reported to the fourth, who reported to the third, who reported to the second, who reported to the first was pretty strange and probably unnecessary. Similarly, if a team of five in a struggling business spent most of their time in meetings deciding which toilet paper to stock in the bathroom, their priorities would seem skewed. Structure can come too early, and cause harm by slowing down a group that should be focused on other things.

However, it's more common in small companies to see structure come too late. The problems creep up slowly. One person gets used to making all of the decisions and changing his mind frequently. This strategy works fine when it's just him and a couple of others. But when he keeps doing it with a team of 10, a team of 20, a team of 50, what you start to see is a high degree of confusion and wasted effort. The cost to change his mind becomes more and more expensive.

One of the best analogies I've heard for startup leadership comes from a friend, On Freud, who's been in engineering management at several different startups. On describes the earliest startup as like driving a race car. You're close to the ground, and you feel every move you make. You have control, you can turn quickly, you feel like things are moving fast. Of course, you're also at risk of crashing at any moment, but you only take yourself down if you do. As you grow, you graduate to a commercial flight. You're farther from the ground, and more people's lives depend on you, so you need to consider your movements more carefully, but you still feel in control and can turn the plane relatively quickly. Finally, you graduate to a spaceship, where you can't make quick moves and the course is set long in advance, but you're capable of going very far and taking tons of people along for the ride.

Assessing Your Role

Recognize the size of the vessel you're steering. This will be determined by a combination of the number of people in the company, the age of the company, the size of the existing business infrastructure (software, processes, and the like), and risk tolerance:

People

The more people you have, the more thoughtful structure you need to get everyone moving in the right direction. Leaders who want a high degree of control over their organization tend to need more structure in place to make sure their wishes are enacted. Modern companies often put their structural focus on goal setting instead of trying to make all decisions from the top, but don't underestimate the structure you need to successfully set and communicate goals.

Age

The longer a company is around, the more habits become entrenched. On the other hand, the longer a company has been around, the more likely it is to continue to survive.

Size of existing infrastructure

If you have few established business rules (such as "this is how we determine what to charge our customers") and little code or physical infrastructure (like stores, warehouses, or inventory), there is less need for structure. On the other hand, the more existing business rules and infrastructure you have, the more you'll need clarity on how to handle them.

Risk tolerance

Are you in a highly regulated industry? Do you have a lot to lose if certain types of mistakes are made? Or are you in an unregulated industry, with little on the line? Your structures and processes should reflect this. In general, the more people you have depending on you and the larger the business is, the less risk you'll be willing to take even without regulatory requirements.

Structure grows as the company grows and ages. In fact, there's even a law that accounts for this, from John Gall's book *Systemantics:*[1]

---

1 John Gall, *Systemantics: How Systems Really Work and Especially How They Fail* (New York: Quadrangle/The New York Times Book Co, 1975).

A complex system that works is invariably found to have evolved from a simple system that worked. A complex system designed from scratch never works and cannot be patched up to make it work. You have to start over with a working simple system.

Your company started as a very simple system that contained a few people, and as more and more people and rules and infrastructure were added, it evolved into a complex system. I don't think there's a huge benefit in overdesigning your team structure or process when your team is small and functioning well. However, at some point you'll start to experience failure, and failure is the best place to investigate and identify where your structure needs to change. In the creating a career ladder example, one person quitting because of a lack of a career path might not be enough to push you to create a career ladder, but you may reconsider when multiple people quit or fail to join. You'll need to weigh the value the lack of structure brings the team against the cost of losing people you might otherwise want to employ.

My advice to leaders is simple: when failures occur, examine all aspects of reality that are contributing to those failures. The patterns you see are opportunities to evolve your structure, either by creating more or different structure or removing it. Think about how often the failure happens and its cost, and use your best judgment about the changes that need to be made. Using failure to guide evolution lets you apply structure at the right level. If a failure is occurring in only one part of the system—say, on one team—you can try to address the structure on that team without necessarily changing the larger structure. What about examining success? Well, you *can* learn things from success, but it is often a poor teacher. Ironically, while luck plays a role in both failure and success, we often attribute failure to bad luck and success to our own actions. As Gall's law says, a simple system that works can evolve into a complex system, but that doesn't mean that applying the lessons from a successful complex system will let you replicate that success in other places. As humans, we tend to blame failure on bad luck until it's impossible to ignore our own contributions to that failure. Therefore, we're less likely to overstructure our teams based on lessons from failure. Success, on the other hand, tempts us with the silver bullet, that one weird trick that could make everything great. If you want to learn from success, make sure you can identify the actual improvement you're seeking when applying those lessons more broadly, and that you understand the context required to repeat that success.

The age of the company and size of the team plays into this issue. If you're at a company that's been around for a while and will be around for a while, using structure (adding or removing) to improve efficiency is very helpful, even if it costs something up front to implement. That's part of the trick. Learning rarely comes for free. Analyzing situations and thinking about good takeaways takes time. If the value of your future time is less than the value of your current time, then you're probably not going to worry too much about saving future time. Just because your company is big, old, and stable doesn't mean you can have as much rigid, unchanging structure as you want. Technology changes often enable formerly risky moves to become safer than the slow-moving alternatives. Software release frequency is a good example of this. For a long time, releasing software frequently was difficult and expensive, largely because you were shipping that software to the user. In the modern SaaS world, bugs can be easily fixed, and the risk involved in shipping a bug is much lower than that of not expanding features quickly enough to keep up with competition. It's this type of unconditional attachment to old structures that makes many people hesitant to adopt structure at all. But if you don't adopt structure when you need it, things can also go wrong.

When every new hire slows the team down for months because there is no onboarding process, that is a failure due to lack of structure. When people regularly leave the company because they have no path to advancement or career growth, that is a failure due to lack of structure. The third time you have a production outage because someone logged directly into the database and accidentally dropped a critical table, that is a failure due to lack of structure. I said earlier that I prefer to talk about learning and transparency rather than using the word *structure*, because really what we're talking about here is identifying the causes of failures, especially frequent failures, and trying to figure out what we can change to solve for those failures. This is fundamentally about learning.

Creating Your Culture

Culture is **how** *things get done, without people having to think about it.*

—FREDERICK LALOUX, *REINVENTING ORGANIZATIONS: A GUIDE TO CREATING ORGANIZATIONS INSPIRED BY THE NEXT STAGE OF HUMAN CONSCIOUSNESS*

Culture is an oft-discussed topic of building startups. What are the core values of the company? What is the company culture like? Are new hires "culture fits"? Is "culture fit" a dogwhistle for discriminatory hiring practices?

One of the things I have come to believe strongly is that culture is real; it's also incredibly important, and it's something that many people don't understand at all. It's both an easy, natural consequence of your company's evolution and something that can quickly become a problem if you don't tend to it. Consciously guiding the culture of your team is part of a leader's job, and to do this well, you need to understand what it means in the first place.

So what is culture? Culture is the generally unspoken shared rules of a community. American culture dictates that we shake hands as a greeting, for example, while in some other cultures, touching strangers is considered very odd. The way you address people of different standings or different relationships to you is part of your culture. Culture doesn't mean that every single person holds exactly the same values, but it tends to guide a general overlap, and it creates a bunch of rules of interaction that you don't have to think much about if you are deeply ingrained in that culture.

People do make decisions using methods other than cultural values. They may adhere to the standards of a formal or informal contract, for example. They may do a pure data-driven analysis and determine the optimal outcome. But in complex environments where the needs of the group must override the needs of the individual, cultural values are the glue that enables us to work as a team and make decisions when faced with uncertainty. This is why figuring out and guiding your culture is such an important part of building a successful company.

If you're forming a new company, there's no guarantee that a predetermined healthy culture will fall out. You may hope that you can create a planned community of people, a community of like-minded individuals who will bind together to create this great workplace and product. But reality is much messier than that. Reality is much more of a race for survival, with culture as an afterthought or a post hoc justification. The early employees will form the culture, for good or for bad—or likely for a mixture of both.

Not every person will fit in at every company. The sooner you realize this, the better. Sometimes we are afraid to have core values because we believe they will create discrimination. I would argue that a thoughtfully created set of values that are actually *values* should reduce the kinds of surface discrimination that often happen at tech companies in favor of creating a real community of employees who share core principles and ways of communication. It is to your advantage to create a culture that allows for bringing a broader range of people into your community. "Engineers who graduated from MIT" is not a culture. "People who value technology innovation, hard work, intellect, scientific process, and data" might be. The first allows only an incredibly narrow subset of humanity to pass through it successfully. The second allows a much broader set of people to fit, while ensuring those people actually have the same values.

If you come into a company with core values, those values were probably created by the founders, or founders and early employees, and thus they reflect the company's culture. This is important to understand, because you'll be measured against these values whether you realize it or not. The founding team's values will be reinforced, recognized, and rewarded inside of the company. My experience has shown that employees who truly embrace and exhibit all of the core values of a company tend to do well naturally. The fit is easy for them. They may get stressed out or work too hard, but they are well liked and usually happy. Those who do not match all of these values as easily will have a harder time. That doesn't mean they will fail, but there will be more friction for them, and it may feel like more work to fit in and feel accepted.

How does this apply to you? If you are a technical executive, cofounder, or CTO, this information has deep applications. If you join or create a company with very different values than your own, you'll feel a great deal of friction that will make your life harder. At the highest levels, all of this cultural alignment comes to play in everything you do, because you spend so much of your time in the land of negotiation, collaboration, and cross-functional teamwork. This doesn't mean that you can't be successful in a company that holds some different values from your own. In fact, it's pretty rare that you agree perfectly with every value of every person on the senior team of a company. You probably don't even agree with every value of every person in your family, or among your friends! Still, the amount of overlap between the traits you value most and the traits your company values most largely determines how easy the fit will be for you.

Applying Core Values

Whether or not you're in a founding or executive position, understanding and cultivating culture is a key part of your job as a leader. Here are some suggestions for how to approach this issue.

First, define your culture. If you have a set of company values, map those values onto your team. You may add a couple of values that are special to your team, or interpret the values in a way that makes sense for your team. On my tech team at Rent the Runway, for example, we explicitly valued diversity. That meant that we were more interested in what you could do and what your potential was than having you fit a certain set of checkboxes in the screening process. We layered a learning culture on top of our company values, because we believed that this was important for us as engineers. The point of this layering is that every subteam will have its own slightly distinct culture. Some teams are focused on being very professional, are in the office for very regular hours, and work in a very regimented way. Some teams prefer later or earlier hours, or less formal meeting cultures, with more room for chatting and hanging out socially.

Second, reinforce your culture by rewarding people for exhibiting its values in positive ways. People can share core value stories at company all-hands meetings. At our technology department all-hands meetings, we would have people give shoutouts to each other for "keeping it dope" and going above and beyond. Some people find this exercise uncomfortable, myself included. Reach through the part of you that is shy about praising people or embarrassed to share your feelings, and go into the part of you that cares about the people you work with. You can share these stories in a way that is not forced or fake. The stories that we tell as a community bond us together.

One of the most important uses of performance reviews is to evaluate the alignment between team members' values and the company's values, and therefore what values should be part of your performance review process. Call out when and how people exhibit some of the core values of the team. This practice reinforces desired behavior in a positive way. It also gives you a sense of who on your team exhibits most or all of the values, and who does not.

Learn to spot people who have values conflicts with the company or team. If your company has a value of "roll up your sleeves and get involved," the teammate who continually pushes off work to others is not truly following this value of the company. If you have a value of "happiness and positivity is a choice," the teammate who pooh-poohs every idea and criticizes everything is going to have problems fitting in. Sometimes, people will change to adopt the values. "Happiness and positivity is a choice" is actually one of the core values of Rent the Runway, and I would not say that I came from a work background of happiness. In fact, I came from a fairly professional and critical work culture. But I learned to appreciate the value of looking at things in a positive light. That doesn't mean that I lost my critical eye, and it was never the easiest value for me to adopt completely, but it wasn't a deal breaker. Using the core values to coach people in areas where they are misaligned can help you articulate what otherwise may feel like just ambiguous friction.

Finally, use this as part of your interview process. Remind your interviewers of the values of the team, and ask them to look out explicitly for places where the interviewee seems to match or collide with these values. A lot of interviews try to determine cultural fit by what I would call "friendship" markers, such as "Would you like being stuck in an airport with this person?" You certainly don't want to hire people that your team can't stand to be around, but cultural fit is not about hiring friends. I've had great working relationships with people that I would not want to chat with for hours outside of work, and terrible working relationships with people I would love to be stuck with in an airport. Furthermore, culture fit as determined by friendship tests is almost certain to be discriminatory in some way. Humans form friendships with people who have significant shared background experiences, and these experiences tend to closely correlate with things like schooling, race, class, and gender. The shortcuts you get by hiring friends are not usually the values you need to form a strong team.

So, don't be vague when discussing fit. Be specific. What are the values of this team, and where have you noticed any match or mismatch? A very smart engineer who really values independence may not be a fit for a team where everyone must collaborate extensively on all projects. Someone who believes that the most analytical argument always wins may not work well in a company that values empathy and intuition over pure analytical skill. I use these examples because all of the values here are compatible in certain situations, and incompatible in others, and that is what makes this a powerful measure. Understand what your company's values are, understand what your team's values are, and think about what you personally value. Write the values down if they aren't already written, and try to be explicit. Use this explicit list to evaluate candidates, praise team members, and inform your performance review process.

Creating Cultural Policy

Creating cultural policy documents can be hard, because getting started on these documents from scratch is hard. Fortunately there are fewer and fewer documents that you need to start from scratch to create, as more people are sharing publicly their policies and processes for everything from career paths to pay scales to incident management. However, just having a starting point and copying it is not always enough. I learned this the hard way when I tried to roll out my first engineering career ladder. As I said earlier in this chapter, there comes a time for adding structure, and that time is usually when things are failing. The failure that drove me to create a career ladder came when our HR team was doing a salary review for the engineering team. I realized that we had no salary structure at all. Because of that lack of structure, most people were paid based on a combination of their previous jobs' salary and their negotiating skills. Additionally, we had a hard time figuring out who we needed to be hiring in. Were we only hiring "senior" engineers? What did that mean? What about management or other roles?

After a nudge from our HR team, I set out to create a ladder (*http://dress code.renttherunway.com/blog/ladder*), which I've cited in pieces throughout the book. I did this by asking my friends who ran other startups if they had one. One of my friends did, and he shared it with me. It had eight levels, from entry-level engineer to executive, broken into four categories: technical skills, getting stuff done, impact, and communication and leadership. I took this ladder, added a few more details, renamed the levels, and rolled it out. This makeshift ladder was very basic. For each level, at each skill, you got one or maybe two sentences on what classified a person as working at that level. Even with some additional information from me, there were perhaps four points you could look at for each category. The worst were the earliest levels, which were the most basic and provided very little guidance to early-career engineers. I delivered the new ladder to my team, and even communicated the new ladder in the same style that my friend used to communicate it to his team. I told them the ladder existed to make sure we were being fair with things like compensation, and it was something they could use to discuss their level with their manager and learn how to grow. I told people it wasn't a big deal, that they shouldn't obsess over their level. I then spent some time talking about John Allspaw's blog post "On Being a Senior Engineer" (*http://www.kitchensoap.com/2012/10/25/on-being-a-senior-engineer/*) in an attempt to inspire the team to push themselves.

Long story short, my first ladder was a flop.

Why did a ladder that seemed to work fine for my friend fail so badly for me? I can only speculate, but there were some pretty big differences between our companies. My company was very diverse in terms of background. I had a team that was mostly pulled from small companies and startups, with a handful of people like myself who had worked at big finance companies and only a couple who had mostly worked for big tech companies. We had no real shared cultural habits to pull from because of this diverse set of work experiences. My friend, on the other hand, managed a team that had a very large, strong core of people who had all worked for the same large tech company, so there was a lot more shared understanding that didn't need to be made so explicit.

I share this story for a very important reason: where my friend was able to succeed, I failed, despite following the same template. This lesson is crucial for anyone who wants to create good team culture. What works for one company—a company that is creating a certain type of product or working in a certain industry—will not always translate well to another company, even if the companies have a lot of things in common. We were both managing startups when we rolled out our respective ladders, and our teams were similar in size, but we needed very different things for our teams to be successful. My first ladder was a flop because my team needed more details. The goal of the lightweight ladder was to keep the team from obsessing over their levels and promotion, but instead the lack of detail caused many of them to obsess even more. Engineers argued that they deserved to be at higher levels because the details were vague. It caused a constant series of headaches.

Writing a Career Ladder

Here are some important issues to consider when writing a career ladder for your organization:

- **Solicit participation from your team.** To write a better ladder, I had to change my approach. First, instead of doing it by myself, I enlisted the support of the senior managers and engineers on the team to provide feedback and details. I asked people to highlight things they didn't understand. I asked them to propose rewrites, additions, edits, and details. We discussed it as a group, and we had subgroups work on the parts of the ladder that they cared about most. For example, the most senior individual contributors worked on the technical and skills expectations for the individual contributor levels.

- **Look for examples.** Second, I got more examples of ladders from friends at other companies to help provide some ideas for the details. There's a lot of good work out there now that you can use if you need to write something, but at the time, I had to do my research by asking people to print out whatever they felt they could share, or give me high-level notes. The best details came from friends at bigger employers, especially those with strong technical reputations. It can be hard to explain the scope of work expected at very senior technical levels, and having those examples from bigger companies really helped us put the details in writing.

- **Be detailed.** One of the biggest challenges you'll face when writing a good ladder is sketching out the details. You want something that is inspirational and descriptive but that matches your company. It doesn't make sense to expect a director for a 50-person engineering team at a startup to manage, say, an entire division, in the same way that it might make sense to have that expectation at a large multinational corporation. Think about the kinds of details you would look for when deciding if someone should be hired in at a level or promoted to a level, and try to include those details as appropriate.

- **Use both long-form descriptions and summaries.** I broke the ladder out into two documents. The first was a shorthand spreadsheet version that allowed me to see the various level attributes side-by-side and see how they evolved through increasing levels. This was helpful because as I wrote, I could see how I built from one level to the next, and how the roles expanded in their scope, skills, and responsibilities. The second document was the long-form version. Writing the long-form version was helpful to me because I felt that I could tell a more complete story about the players at each level. Instead of just visualizing the level as a set of skills and attributes, the long-form ladder reads a bit like a performance review of a person operating well at each level. You—and your employees—can see how those skills work together to form a complete role. How many levels should your ladder have? You'll need to answer two more questions to figure this one out. First, how do you pay people? And second, how do you recognize achievement?

- **Consider how the ladder relates to salary.** Your HR department will want to use the career ladder to help set salary expectations. Usually, each level will have a salary band, or a range between a minimum and maximum base salary that a person in the level can earn. If you don't have many levels, you'll need to have very wide salary bands to account for the fact that two people within that level can perform very differently, and to account for the fact that engineers tend to expect to get salary increases frequently, especially in the earlier parts of their careers.

- **Provide many early opportunities for advancement.** Some people advise having a lot of levels toward the beginning of the ladder to account for the fact that early-career engineers expect frequent raises and promotions. You may want to be able to promote someone every year for the first two to three years of her career. If that's the case, create several levels that encompass the role of *software engineer* and provide relatively narrow salary bands for those levels, on the expectation that people in those roles are either being promoted quickly or moving on from your company.

- **Use narrow salary bands for early-career stages.** Lots of levels and narrow salary bands mean that you can promote people quickly and justify giving them raises while keeping your pay for all people at a certain level close to the same. This is good if you are worried about paying fairly and avoiding bias that might lead you to, say, pay men more than women at the same level. Unfortunately, it's incredibly hard to create enough detail between close levels to allow a person to easily distinguish someone being at one level or another.

- **Use wide salary bands when and where you have fewer levels.** Wide salary bands and few levels make a clearer distinction between the skills at each level, and should make it easier to tell who is operating at which level. In the case of widely spaced levels, you want to have large salary bands and you want those salary bands to overlap. So, a software engineer band may go $50–100K, and a senior software engineer band may go $80–150K. That means a strong software engineer may make more than a senior engineer. You need this wiggle room to retain talent who are performing well at their current levels but are not ready to take on the additional responsibility of the next level. You will also find yourself using this wiggle room to hire people who are on the fence into the lower level with the expectation that they will be promoted quickly.

- **Consider your breakpoint levels.** It is common for companies to have certain levels that they consider "up or out." These are early-career roles where a lack of advancement means that the person has not achieved the maturity or independence needed to remain at the company. This policy tends to get translated into your ladder as an implicit or explicit breakpoint level. What is the lowest level at which people can sit forever, never getting promoted but also not underperforming? This is your breakpoint level. For many companies, it's somewhere around senior engineer. Someone who's made it this far is a solid team member, but he may stay at this level indefinitely by his own choice. It's good to have a notion of where this is. You may even want to use it as the point at which your ladder levels get harder to achieve. Expect your team to cluster around this level, with fewer people above or below it.

- **Recognize achievement.** Some companies want to keep levels secret, but that tends to be impossible. People will talk. However, you can go out of your way to emphasize certain levels while keeping other levels secret, possibly even from the employees themselves. Some HR departments have separate pay grade numbers that they use to track employee pay that are disconnected from career ladders entirely. I am not advocating for this. However, I do encourage you to have at least some of your levels be keystone promotions, which are shared and celebrated. I think that the promotion to senior engineer is a big deal, as well as the promotion to staff engineer and, if you have such a role, principal engineer. On the management track, a promotion to director is worth celebrating, as is a promotion to VP. Having keystone levels that are not too close together gives people a bigger achievement to strive for beyond the next pay increase, and keeps these levels feeling important from a larger career standpoint.

- **Split management and technical tracks.** It's pretty obvious in this day and age that you need separate tracks for management and individual contribution. You do not want people to feel that the only path to advancement is by managing people, because not everyone is suited to that role. Commonly, you'll see a split above senior engineer where organizations start to specify management levels and technical levels. However, you should not necessarily expect to have the same number of people in the senior technical levels as you do in the senior management levels. Senior management is generally a volume-driven need. You need enough managers to manage the people you have on the team. Senior technical depends on the complexity and scope of technical leadership that your teams and products require. It is possible to have a large team with few senior technical people, or a small team with many senior technical people and fewer managers. It would be unusual to have a perfect balance here.

- **Consider making people management skills a mid-career requirement.** Encourage everyone to have some sort of management or mentorship experience before they are eligible to be promoted above the level of the track split. For most companies, the tracks should split when people start to exhibit leadership, whether that leadership involves managing humans or designing software. But even when designing software, you're dealing with other humans and human needs. Great senior individual contributors still know how to manage projects and mentor more junior members of their team, so consider making leadership experience (usually via acting as a tech lead) a requirement for promotion to senior individual contributor levels.

- **Years of experience.** No one likes to put artificial barriers onto people, and years of experience can feel like the most artificial barrier. With that said, I encourage you to be wise on this issue. In my ladders, I distinguish the keystone levels by an expectation of maturity increases, and these tend to correspond with years of experience in the industry as well as, to a lesser extent, age. For example, take the case of staff engineer. It takes a lot of individual maturity to think through large projects, which is, in my view, the distinguishing feature of a staff engineer. Being a brilliant programmer is not enough to be a great staff engineer; you need to have shown a track record of completing and supporting some long-running work to justify this title. You don't have to put years of experience as a strict requirement for levels, but consider having some rules of thumb, especially if you are writing a ladder for the first time and rolling out levels.

- **Don't be afraid to evolve over time.** When you write a ladder like this, you're creating a living document that will need to evolve as your company grows. You're probably going to miss some details. My ladder was hard for frontend-focused developers to interpret because of my own focus on infrastructure development, so we needed to tweak it to better account for what it meant to be a senior performer within that world.

A good ladder is a critical element to use in hiring, in writing performance reviews, and of course in the promotion process. If you have the chance to create such a document, don't be afraid to involve your team. The best processes and documents reflect the team as a whole and not just your bias at the moment, and one of the greatest things about setting these ladders up in a small company is that you can involve a lot of people without a ton of bureaucracy in the process.

Cross-Functional Teams

Who do you work with? Who do you report to? Who do you collaborate with? The answers are obvious in both extremely small companies (answer: everyone) and very large companies (answer: there's a pretty clear structure that was set up before you joined). As a leader at a growing company, you'll need to help answer these questions for your team and your company at least once, and probably multiple times. What should the answers be?

I want to take some time to talk about one of the best things I experienced in my work at Rent the Runway: the evolution of our product engineering organization. When I joined, the engineering team was divided into roughly two groups: *storefront*, which did all development for the customer-facing website, and *warehouse*, which supported the software that ran the warehouse operations. We quickly evolved storefront to be frontend and backend because we were rewriting the code from a PHP monolith to a Java- and Ruby-based microservices architecture.

Toward the end of my first year, we ran an experiment. We had a new product we wanted to build for the customer, a feature based on our customer photo reviews. Because finding a dress that would fit well was a challenge for our customers, we wanted to enable shoppers to see photos that other customers had uploaded showing themselves in the dresses, along with customer-provided information about their normal size, height, weight, and "shape" (athletic, pear, curvy, etc.). To implement this feature, we created a cross-functional team. We had engineers who specialized in the frontend user experience development, and engineers who worked on the backend services. We had a product manager, designers, a data analyst, and even a representative from the customer service team. This cross-functional team worked as a group to design and deliver this feature to our customers.

This project was a massive success. We delivered a good feature, fairly quickly, and the contributors all felt that they understood the goals of the project and were able to work better because of this cross-functional team. Prior to this project we had been deep in a pattern of "us versus them," where your particular business function was "us" (tech, product, analytics, marketing, etc.), and the rest of the organization was "them." Creating these collaboration units gave people a chance to see the whole group as "us." It was a clear win in terms of organizational health, so we evolved our whole organization to have all product engineering performed by cross-functional teams like this. Call them what you want— "pods" or "squads" or "pillars"—but cross-functional product development groups are a popular structure for a good reason. By putting everyone who is needed to make a project successful together in one group, you help the members of those teams focus on the project at hand, and you make the communication for the whole group much more effective.

Conway's Law (*http://www.melconway.com/Home/Conways_Law.html*) is often cited in discussions of this kind of structure. It states: "Organizations which design systems...are constrained to produce designs which are copies of the communication structures of these organizations."

When we put cross-functional teams together, we are acknowledging that the most important communication—the communication that we need to favor above all else—is that which leads to effective product development and iteration. Note that this structure will not necessarily produce the most effective technology! In fact, it will probably produce systems that have some inefficiencies compared to companies that have a more engineering-centered team structure. So, should you adopt this structure, you have to decide where you're willing to take some system design hits in order to most effectively create products.

STRUCTURING CROSS-FUNCTIONAL TEAMS

How do the nuts and bolts of such "pod" structures work? One element that often causes anxiety is who is managing whom. When we moved to this team organization, we did not change the management structures. Engineers were managed by engineering managers and reported in to me. Product managers reported to the head of product. But determination of who was working on what was done largely by the pod itself. This meant that you could still get technical guidance and oversight from your engineering manager, but your day-to-day work was determined by the needs of the pod's roadmap.

Of course, every function has its own focused needs. Usually someone in engineering needs to oversee critical core systems, and you probably need a few specialists around for things like the core web platform, mobile, or data engineering. I kept these functions in a small infrastructure organization that was not generally assigned to product development. Even with a dedicated infrastructure group, the engineers assigned to product pods still need some time to account for engineering-specific tasks like on-call, interviewing, and sustaining engineering (aka technical debt). I advise reserving 20% of all engineering time for such work, based purely on my personal experience and the experience of my peers in engineering management.

This cross-functional structure is not unique to small startups. Many large companies also structure their teams in this fashion. Banks, for example, often have technology teams that are attached to specific areas of the business, and while the management structure is formed by engineers, the roadmap and day-to-day work are jointly determined by the needs of the business unit and its associated engineering team. There is generally a centralized infrastructure team that supports both fundamental systems as well as large frameworks and technologies that will be used by many teams across the company. Even many technology companies are structured in this way, although the "business units" may themselves be headed by former engineers who act as product or business managers instead of business specialists.

The implications of the cross-functional structure are subtle. The values of everyone in these teams will start to change. In technology-focused structures where engineers work solely with other engineers, particularly engineers of their same "type" (mobile, backend, middleware, etc.), the focus is on being the best engineer by some measure of engineering excellence. People who design complex systems or who know the details of the latest iOS are the leaders and role models for the teams. In a product-focused structure, the leadership focus changes. Now the engineers who have the best product sense, the engineers who are capable of getting features done quickly and efficiently, and the engineers who communicate the best with the other functions will start to emerge as the leaders of the team.

I mean no value judgment here, but I encourage you to be aware of the product/business versus technology focus and apply it where it makes sense. What is truly important to the success of your company or your organization? If the most important thing is evolving a product that is a function of many different business areas coming together, you probably want leaders who have that business sense. On the other hand, in the areas where the technology must be rock-solid or exceptionally innovative and cutting-edge, you probably want teams that have more of an engineering focus and that are led by people who can design complex systems. You don't have to go entirely one way or the other, but recognize that one of these will lead the company as a whole, and—especially if your role is in senior management—focus your skill set on the one that the company itself most values and hire in for the other.

Developing Engineering Processes

I've had to deal with many different engineering processes over the years. I remember the first time I worked in a code base with unit tests that we were expected to run before checking in code. I was very diligent about doing this, and very upset every time someone broke the build because she hadn't bothered to make sure her change didn't break the tests. I also remember the first time I had an engineering process forced upon me that I hated. After years of no required code reviews, no ticketing, and no tracking, a central bureaucracy decided that everyone had to adopt all of these measures at once in a push for standard software development lifecycle management. It felt unnecessary, slow, and burdensome, and no one bothered to explain to us why these changes were happening.

Engineering processes are the place where the rubber meets the road when it comes to structure. Career ladders, values, team structures—all of those are easy compared to the general angst and frustration that you can cause by adopting the wrong engineering processes for your teams. Without any process, your teams will struggle to scale. With the wrong process, they will be slowed down. Balancing the current size and risk tolerance of your team with the processes at hand is the essence of guiding good software development and operational guidelines.

Ask the CTO: Engineering Process

I'm the head of engineering at a small but quickly growing startup. We have very little process right now: there are no code reviews, we use Trello to manage tasks but rarely put everything into that system, and our architecture decisions tend to be made by whomever is working on the project at the time, with my sign-off.

Recently, some engineers have come to me to complain that new people are checking bad code into the systems. They want us to introduce code reviews for all changes. I also just discovered that someone has been writing a new system in Scala, despite the fact that all of the rest of our code is in Ruby. He's the only person who knows Scala on the team, and I'm afraid of the support burden, but the project is pretty far along so I can't just kill it.

What should I do? I'm nervous about going from zero process to a bunch of process all at once, but something needs to change!

Think of process as risk management.

As your teams and systems grow, it's almost impossible for any one person to keep the systems in her head. Because we have a bunch of people coordinating work, we evolve processes around that work coordination in order to make risks obvious.

One way to think about engineering processes is that they serve as a proxy for how hard or rare it needs to be for something to happen. A complicated process should exist only for activities that you expect to be rare, or activities where the risks are not obvious to people. "Complicated" in this case does not only mean a long process. Sometimes, the complexity is in getting sign-off from a group of people who are very busy, or in meeting a very high standard.

This has two important implications. The first is that you should not put a complicated process on any activity where you want people to move quickly and where you believe the risk for change in that activity is low or that the risks themselves are obvious to the whole team. If you want to do code review for all changes, make sure that the process for code review is not so onerous that the team slows down significantly on minor changes, because that will impact your whole group's productivity.

The second implication is that you need to be on the lookout for places where there is hidden risk, and draw those hidden risks out into the open. There's a saying in politics that "a good political idea is one that works well in half-baked form," and the same goes for engineering processes. The processes should have value even when they are not followed perfectly, and that value should largely lie in the act of socializing change or risk to the team as a whole.

Practical Advice: Depersonalize Decision Making

There are three major processes that you should consider adding as your team grows. All of these processes work best when you set behavioral expectations around them in addition to the technical details.

CODE REVIEW

Code review is, for better or worse, a modern standard. Once you have a team of a certain size with a certain number of people working on a code base, code review can be a valuable tool for ensuring the stability and long-term quality of that code base. However, required code review will also be on the critical path of getting work done, so you want the process to be straightforward and efficient. Additionally, code review is often a place where engineers behave poorly toward one another, using it as a platform to criticize their colleagues or to enforce unrealistic standards. Here are a few best practices to smooth the way:

- **Be clear about code review expectations.** For the most part, code reviews don't catch bugs; tests catch bugs. The exceptions to this rule are that code review can catch missing updates to comments or documentation or missing changes to related features, and code reviewers can sometimes tell when there is inadequate testing of the new or changed code. Code review is largely a socialization exercise, so that multiple team members have seen and are aware of the changed code.

- **Use a linter for style issues.** Engineers can waste absurd amounts of time on questions of style, specifically formatting. This should not be up for debate in code review. Decide on a style, and put that style into a linter that formats the code automatically. Allowing style to be up for discussion in code review often leads to nitpicking and criticism that can feel unproductive at best, and bullying at worst.

- **Keep an eye on the review backlog.** Some companies implement a limit on how many outstanding review requests a person can have assigned to him, and they block that person from requesting review when he has too many outstanding requests. Think about how you want to get these requests pushed through the system, and how you can make sure that everyone gets adequate time in their code.

THE OUTAGE POSTMORTEM

I'm not going to talk about the details of incident management, but the "postmortem" process is a critical element of good engineering. In fact, instead of calling the process a postmortem, many have started calling it a "learning review" to indicate that its purpose is not determining cause of death but learning from the incident. There has been a lot written on this topic, so I'll highlight only a few elements that I believe are critical, especially for small teams:

- **Resist the urge to point fingers and blame.** It is incredibly tempting, after a stressful outage, to point fingers and ask people why they failed to foresee the consequences of their behavior. Why did they run that command on that box? Why didn't they test that? Why did they ignore that alert? Unfortunately, this blaming only results in people being afraid to make mistakes.

- **Look at the circumstances around the incident and understand the context of the events.** You want to understand and identify the factors that contributed to this incident. This might include looking for tests that would have detected the problem, or tools that could have made the incident management go more smoothly. Getting a good list of these circumstantial contributors helps you detect patterns or areas for improvement, and forms the "learning" part of the learning review.

- **Be realistic about which takeaways are important and which are worth dropping.** Be careful not to give the impression that people need to solve every problem they identify in the course of the exercise. Many learning reviews end in a laundry list of things that could be improved—everything from cleaning up alerts to adding role restrictions to following up with a third-party vendor to understand its API. It's unlikely you will get to all of these, and in fact, it's likely that if you try to do all of them, you will end up doing none of them. Pick the one or two that are truly high-risk and highly likely to cause future problems, and acknowledge the ones that you are going to let go for now.

ARCHITECTURE REVIEW

I'm going to roll into architecture review all major systems and tools changes that the team may wish to make. The goal of architecture review is to help socialize big changes to the appropriate group, and to make the risks for those changes clear. Some questions that you may ask people to come prepared to answer include:

- How many people on the team are comfortable using this new system/ writing this new language?

- Do we have production standards in place for this new thing?

- What is the process for rolling this out and training people to use it?

- Are there new operational considerations for using this?

And here are some guidelines:

- **Be specific about the kinds of changes that need architecture review.** Usually these include new languages, new frameworks, new storage systems, and new developer tooling. People often want to have architecture review to prevent teams from designing new features poorly, but it is usually unrealistic to try to catch new feature design early enough in a small company, and it's hard even in a large company. It also slows things down a lot, and as to our earlier point, you probably don't want to put a heavy process in front of a common activity like feature design.

- **The value of architecture review is in preparing for the review.** Asking for review of big changes or additions to the systems forces people to think about why they want to make these changes. Again, one value of these processes is to help make people aware of risks that they may not have considered. You may or may not choose to have the team answer the question of why you should make this change in the first place. I have found that when people are willing and able to get through the requirements for making the change at all, the why is obvious.

- **Choose the review board wisely.** You want the review board to include the people who will be most affected by the change, not just a static chosen group of gurus. Part of the goal is to get yourself out of the hot seat for making every technical decision, and part of the goal is to make sure that those who will need to deal with the outcome of the decision are part of evaluating it. You want these decisions to consider the wider team, and for the wider team to be bought in on them. There is no reason this needs to be company-wide. The scope of the deciding group is best kept to the people who will be closely impacted by the decision. There's nothing more demoralizing than having someone from a completely unrelated area veto a project.

Assessing Your Own Experience

- What policies do you have now? What practices? Have you written any of them down yet? When was the last time you revisited them?

- Do you have company values? What are they? How do you recognize them in your team?

- Do you have a career ladder? Do you feel it accurately reflects the team today? Does it reflect the team you want to have in the future? If not, can you improve it?

- What risks are most concerning for your team? For your company? How can you mitigate those risks without burdening your team with unnecessary processes and bureaucracy?

Conclusion

And so, here we are. You've walked with me down the path from mentor to manager to senior leader. Along the way I hope that you've learned a few tricks, identified a few pitfalls to watch out for, and felt inspired to meet the challenge of whatever role you're in.

The most important lesson I've learned is that you have to be able to manage yourself if you want to be good at managing others. The more time you spend understanding yourself, the way you react, the things that inspire you, and the things that drive you crazy, the better off you will be.

Great managers are masters of working through conflict. Getting good at working through conflict means getting good at taking your ego out of the conversation. To find a clear view of a complex situation, you must see past your interpretations and the stories you're telling yourself. If you want to be able to tell people hard things and have them hear what you have to say, you must be able to tell them without embellishing the facts with your storyline. People who seek out management roles often have strong views on how things should be. That decisiveness is a good quality, but it can hinder you when you fail to see your interpretation of a situation is just that: an interpretation.

Learning to recognize the voice of your ego is one of the benefits of meditation, and when I wrote the first draft of this book it included a series of meditations at each level. For me, having a meditation practice has been essential to developing self-management and self-awareness. Meditation isn't a cure-all, but it can be a useful exercise to practice that awareness of your own reactions, and for that reason I recommend trying it for a while if you are interested. Some of my favorite resources include the podcasts on *tarabrach.com* and the writings of Pema Chödrön.

One other trick I use to get away from my ego is curiosity. I also have a daily habit of writing a page or two of free-flow thoughts every morning, to clear my mind and prepare for the day. I always end with the mantra "Get curious." For me, becoming a great leader was a series of difficult lessons, mistakes, and challenges. Nothing about it was easy, and I was often frustrated with the interpersonal situations I found myself in. Inevitably, when I told my coach about these situations, she would advise me to think about things from the other person's perspective. What are they trying to do? What do they value? What do they want and need? Her advice, always, was to get curious.

So I leave you with that thought. Look for the other side of the story. Think about the other perspectives at play. Investigate your emotional reactions, and observe when those reactions make it hard to see clearly what's going on around you, what needs to be said. Apply that curiosity to people. Apply it to process. Apply it to technology, and strategy, and business. Ask questions, and be willing to have your notions proven wrong.

Stay curious, and good luck on your path!

Index

About the Author

Camille Fournier is an experienced leader with the unique combination of deep technical expertise, executive leadership, and engineering management.

Colophon

The image on the cover of *The Manager's Path* is an illustration by Edie Freedman and Michael Oréal.

The cover fonts are URW Typewriter and Guardian Sans. The text font is Adobe Minion Pro; the heading font is Adobe Myriad Condensed; and the code font is Dalton Maag's Ubuntu Mono.

Learn from experts.
Find the answers you need.

Sign up for a **10-day free trial** to get **unlimited access** to all of the content on Safari, including Learning Paths, interactive tutorials, and curated playlists that draw from thousands of ebooks and training videos on a wide range of topics, including data, design, DevOps, management, business—and much more.

Start your free trial at:
oreilly.com/safari

(No credit card required.)